Student Solutions Manual for Devore/Peck's

Statistics

The Exploration and Analysis of Data

Fourth Edition

Pam Iyer

DUXBURY

THOMSON LEARNING

Australia • Canada • Mexico • Singapore • Spain • United Kingdom • United States

DUXBURY

™

THOMSON LEARNING

Assistant Editor: *Seema Atwal*

Marketing Manager: *Tom Ziolkowski*

Marketing Assistant: *Ericka Thompson*

Editorial Assistant: *Ann Day*

Production Coordinator: *Dorothy Bell*

Cover Design: *Cassandra Chu*

Print Buyer: *Micky Lawler*

Printing and Binding: *Webcom Limited*

For more information about this or any other Duxbury product, contact:
DUXBURY
511 Forest Lodge Road
Pacific Grove, CA 93950 USA
www.duxbury.com
1-800-423-0563 (Thomson Learning Academic Resource Center)

For permission to use material from this work, contact us by
Web: www.thomsonrights.com
fax: 1-800-730-2215
phone: 1-800-730-2214

Printed in Canada

10   9   8   7   6   5   4   3

ISBN 0-534-38043-3

TABLE OF CONTENTS

Chapter 1
The Role of Statistics

Section 1.2

1.1   Descriptive statistics is made up of those methods whose purpose is to organize and summarize a data set. Inferential statistics refers to those procedures or techniques whose purpose is to generalize or make an inference about the population based on the information in the sample.

1.3   The population of interest is the entire student body (the 15,000 students). The sample consists of the 200 students interviewed.

1.5   The population consists of all single-family homes in Northridge. The sample consists of the 100 homes selected for inspection.

1.7   The population consists of all 5000 bricks in the lot. The sample consists of the 100 bricks selected for inspection.

Chapter 2
The Data Analysis Process and Collecting Data Sensibly

Section 2.1

2.1	a	numerical (discrete)
	b	categorical
	c	numerical (continuous)
	d	numerical (continuous)
	e	categorical

2.3	a	discrete
	b	continuous
	c	discrete
	d	discrete
	e	continuous
	f	continuous
	g	continuous
	h	discrete

Section 2.3

2.5   Pregnant women are interviewed and asked if they suffered severe morning sickness early in pregnancy. The baby's sex is noted.
This scientific study is an observational study because we are interested in answering questions about characteristics of an existing population. The condition of morning sickness could not be controlled by the experimenter.

2.7   The article does not report if SUV owners wear seat belts. It states that 98% of those injured or killed were not wearing seat belts. It may be that most SUV owners do not wear seat belts, which would explain the high rate of deaths by ejection.

2.9   It is possible that other confounding variables may be affecting the study's conclusion. It could be that men who eat more cruciferous vegetables are also making a conscious choice about eating healthier foods. A definitive causal connection cannot be made based on an observational study alone.

Section 2.4

2.11 Since each case has a case number, write each case number on a slip of paper; one case number per slip of paper. Place the 870 slips of paper into a container and thoroughly mix the slips. Then, select 50 slips (one at a time, without replacement) from the container. The 50 cases whose case numbers are on the 50 selected slips constitute the random sample of 50 cases.

2.13 Stratified sampling would be worthwhile if, within the resulting strata, the elements are more homogeneous than the population as a whole.

    a   As one's class standing increases, the courses become more technical and therefore the books become more costly. Also, there might be fewer used books available. Therefore, the amount spent by freshmen might be more homogeneous than the population as a whole. The same statement would hold for sophomores, juniors, seniors, and graduate students. Therefore, stratifying would be worthwhile.

    b   The cost of books is definitely dependent on the field of study. The cost for engineering students would be more homogeneous than the general college population. A similar statement could be made for other majors. Therefore, stratifying would be worthwhile.

    c   There is no reason to believe that the amount spent on books is connected to the first letter in the last name of a student. Therefore, it is doubtful that stratifying would be worthwhile.

2.15 Both procedures are unbiased procedures. It would be helpful to consider any known patterns in the variability among trees. Different rows may be exposed to different degrees of light, water, insects, nutrients, etc. which may affect the sugar content of the fruits. If this is true, then Researcher A's method may not produce a sample that is representative of the population. However, if the rows and trees are homogeneous, then the convenience and ease of implementation of Researcher A's method should be considered.

2.17   **a**   The population is all 16-24 year olds that live in New York state but not in New York City.

    b   Since the sample selection excluded New York City residents, we should not generalize to this group. Since fewer New York City young adults drive, the behavior of individuals who live in New York City could differ in important ways from that of individuals in the population sampled.

    c   Since the sample selection included only New York residents and only those aged 16-24, it would be unwise to generalize to other age groups or locations.

2.19 Firstly, the sample consists of only women and male responses may be different from the women's responses. Secondly, the participants are all volunteers and volunteer responses usually differ from those who choose not to participate. And thirdly, the participants are all from the same university which may not be representative of the entire nationwide college population.

2.21   This is an example of selection bias. The example states that 26% of Americans have Internet access. This means that a large portion (74%) of the population have no chance of being included in a public opinion poll. If those who are excluded from the sampling process differ from those who are included, with respect to the information that is being collected, then the sample will not be representative of the population.

Section 2.5

2.23   **a**   Strength of binding.

b   Type of glue.

c   Number of pages in the book and whether the book is paperback or hardback. Other factors might include type of paper (rough or glossy), proportion of paper made with recycled paper, and whether there are tear-out pages included in the book.

2.25   **a**   There are several extraneous variables, which could affect the results of the study. Two of these are subject variability and trainer variability. The researcher attempted to hold these variables constant by choosing men of about the same age, weight, body mass and physical strength and by using the same trainer for both groups. The researcher also included replication in the study. Ten men received the creatine supplement and 9 received the fake treatment. Although the article does not say, we hope that the subjects were randomly divided between the 2 treatments.

b   It is possible that the men might train differently if they knew whether they were receiving creatine or the placebo. The men who received creatine might have a tendency to work harder at increasing fat-free mass. So it was necessary to conduct the study as a blinded study.

c   If the investigator only measured the gain in fat-free mass and was not involved in the experiment in any other way, then it would not be necessary to make this a double blind experiment. However, if the investigator had contact with the subjects or the trainer, then it would be a good idea for this to be a double blind experiment. It would be particularly important that the trainer was unaware of the treatments assigned to the subjects.

2.27   Let us evaluate this design by considering each of the basic concepts of designing an experiment.

Replication: Each of the 8 actors was watched on tape by many of the primary care doctors.
Direct Control: The actors wore identical gowns, used identical gestures, were taped from the same position and used identical scripts.
Blocking: not used
Randomization: The article does not indicate if the 720 doctors were randomly divided into 8 groups of 90 doctors and each group randomly assigned to watch one of the actors on tape, but it is reasonable to assume this was done.

This design appears to be good because it employs many of the key concepts in designing an experiment. One possible improvement would be to randomly select the 720 primary care doctors from the entire population of primary care doctors. By randomly selecting a sample from the entire population, we can generalize our results of the study to the whole population. In this study, the conclusions only apply to this group of 720 doctors.

2.29   Yes, blocking on gender is useful for this study because 'Rate of Talk' is likely to be different for males and females.

2.31   Since all the conditions under which the experiment was performed are not given in this problem, it is possible that there are confounding factors in the experiment. Such factors might be the availability of cigarettes, the odor of cigarettes in the air, the presence of ashtrays, the availability of food, or magazines in the room that contain cigarette ads. Any of these factors could explain the craving for cigarettes. Assuming that the researchers were careful enough to control for these extraneous factors, the conclusion of the study would appear to be valid.

Supplementary Exercises

2.33   **a**   By stratifying by province, information can be obtained about individual provinces as well as the whole country of Canada. Also, alcohol consumption may differ by province, just like we expect differences among states in the US.

b   Occupation is one socioeconomic factor that could be used for stratification. Alcohol consumption habits may be different based on a person's job. For example, a corporate businessman is likely to have more corporate sponsored social events involving alcohol consumption than a day care worker. Yearly income is another factor to use for stratification. Since alcoholic drinks are not free, those people with a high yearly income are likely to be able to afford the alcoholic drink of their choice.

2.35   The manager should number each room. For a simple random sample, he should write the number of each room on separate sheets of paper, put all the numbered papers into a hat and randomly select 15 papers. Those are the numbers for the 15 rooms he should survey. For a stratified sample, he should again write the number of each room on separate pieces of paper, and then divide the papers into piles entitled economy, business class and suites. Randomly choose 5 rooms from each grouping.

A stratified random sample would be more appropriate because the quality of housekeeping may vary from one room type to the next, but remain fairly similar within a given type. Possible extraneous variables include differences in cleaning style among housekeepers, room arrangements (some rooms may be harder to clean than others), and hotel occupancy rate (housekeepers may be rushed when the hotel is full).

2.37   Divide the 500m square plot into 4 equal size subplots, each measuring 250m x 250m, using two rows and two columns. Now divide each subplot again into 4 equal size smaller plots, each measuring 125m x 125m, using the same pattern. The result is, the 500m square plot is divided into 16 subplots with 4 rows of 4 subplots in each row. Now arrange the 4 types of grasslands so that each type appears in every row and column and in every 2x2 subplot. This is done to allow for repetition for each treatment (different grasslands). List all possible arrangements such that these conditions are held, and randomly select one to use in the experiment. Randomization is used in selecting the type of grassland arrangement for the plot as an effective way to even out the influences of extraneous variables. A few of the possible confounding variables in this experiment include exposure to sun or shade, proximity to water, slope of the land or possibly the number of worms in the soil. This study is an experiment since we are observing how a response variable (nesting) behaves when one or more factors (grasslands) are changed.

2.39   There are many possible designs any of which would be a reasonable design. Here is one example. Assume that the paper is published six days a week and assume that the lender is willing to advertise for six weeks. For the first week randomly select two days of the week on which advertisement one will be run. Then, select randomly two days from the remaining four on which advertisement two will be run, and run advertisement three on the remaining two days. Repeat this randomization process for each week of the experiment. If the newspaper has two sections in which an advertisement can be placed, then randomly select three of the weeks and place the advertisement in section one, with the advertisement being run in section two during the remaining three weeks. The randomizations described should control the extraneous factors such as day of week, section of paper, and daily fluctuations of interest rates.

Chapter 3
Graphical Methods for Describing Data

Section 3.1

3.1 **a**

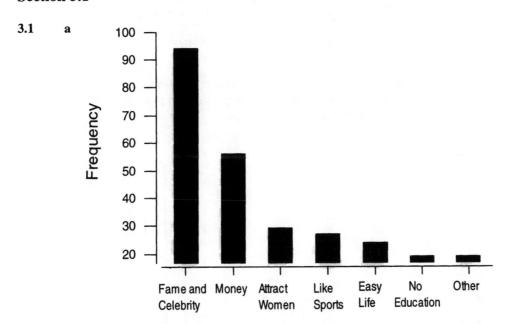

b   A pie chart would also be a good way to display the data. There are only seven categories and it is easy to calculate relative frequencies.

3.3

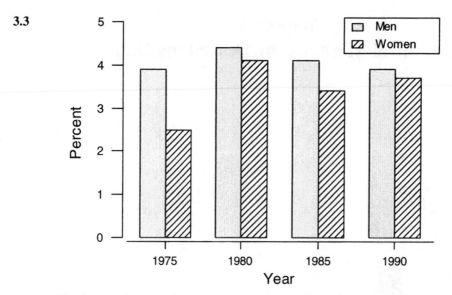

The bar graph shows that a greater proportion of men have medical career aspirations than women. However, over the years, the difference between these two proportions is has decreased. In 1990, the two proportions are almost equal.

3.5

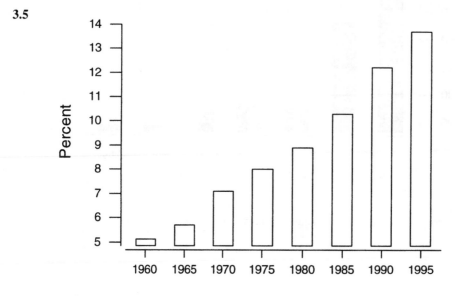

The bar graph shows a steady increase in the US gross domestic product spent on health care over the 1960-1995 time period. The largest increases were in the most recent years, 1990 and 1995.

3.7

Pie Chart of Usage

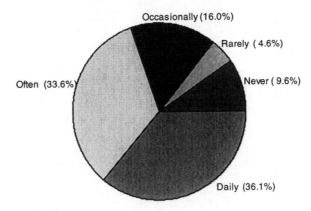

A large proportion of the doctors selected for this study are Internet users. Approximately 70% of the doctors in this survey use the Internet at least once a week. Only about 14% responded 'Rarely' or 'Never'.

3.9   **a**

Day of Week	Frequency
Sunday	109
Monday	73
Tuesday	97
Wednesday	95
Thursday	83
Friday	107
Saturday	100
	n = 664

b $\dfrac{(107+100+109)}{664} = \dfrac{316}{664} = .4759$, which converts to 47.59%.

c If a murder were no more likely to be committed on some days than on other days, the proportion of murders on a specific day would be $1/7 = .1429$. So, for three days the proportion would be $3(.1429) = .4287$. Since the proportion for the weekend is .4759, there is some evidence to suggest that a murder is more likely to be committed on a weekend day than on a non-weekend day.

3.11   **a**

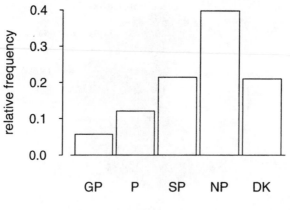

Opinion of Athletes with GPA of 3.0 - 4.0

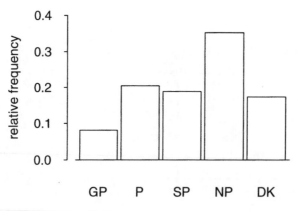

Opinion of Athletes with GPA of 2.0 -< 3.0

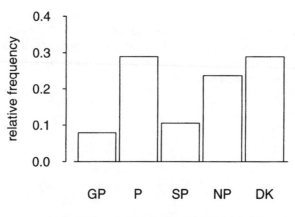

Opinion of Athletes with GPA of 1.0 -< 2.0

The opinions of athletes in the two groups 3.0 − 4.0 and 2.0 −< 3.0 appear similar. The relative frequencies for the 5 categories are about the same for these two groups. However, the opinions of athletes with GPA's of 1.0 −< 2.0 are quite different from the other two groups, which is reflected in the histogram for the 1.0 −<2.0 group. For instance, about 29% of the 1.0 −<2.0 group responded "don't know", but for the other two groups about 21 % and 17% answered "don't know". About 29% of the 1.0 −< 2.0 group answered "a problem", while for the other two groups only 12% and 20% answered "a problem" . About 24% of the 1.0 −< 2.0 group answered "not a problem", while for the other two groups about 40% and 35% answered "not a problem". The primary difference is in how the three groups answered "a problem" and "some problem". This is more easily seen in the graph that appears in part **b**.

b

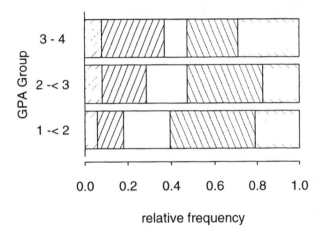

(See the discussion in part **a**.)

Section 3.2

3.13   Your dot plot should look similar to the one given in the text. Slight differences might appear because of the accuracy of placement of the dots. The following dot plot was created using MINITAB with a requested width of 49 characters.

```
          .      ..    ....    :..      .  .:. .  .....       .      . .
      -----+---------+---------+---------+---------+---------+---Percent
          16.0       24.0      32.0      40.0      48.0
```

3.15   **a**   2.2 liters/min.

b   In the row with stem of 8. The leaf of 9 would be placed to the right of the other leaves.

c   A large number of flow rates are between 6.0 and 8.0. Perhaps 6.9 or 7.0 could be selected as a typical flow rate.

d   There appears to be quite a bit of variability in the flow rates. While there are a large number of flow rates in the 6.0 to 9.0 range, the flow rates appear to vary quite a bit in relation to one another.

e   The distribution is not symmetric. Taking 7.0 as a typical value, the smaller flow rates are spread from 2.2 to 7.0, while the larger flow rates are spread from 7.0 to 18.9 (a larger spread).

f   The value 18.9 appears to be somewhat removed from the rest of the data and hence is an outlier.

3.17

```
1H | 8
2L | 0 2 4
2H | 9 8 7 5 6 9 7 8 5 8 5 5 8 7 7 8 7 9 6 6 7 5 8 5
3L | 1 4
3H | 6                      stem: ones
4L |                        leaf: tenths
4H | 7 9
5L | 4
```

3.19

```
64 | 64 70 35 33          stem: hundreds
65 | 26 83 06 27          leaf: ones
66 | 14 05 94
67 | 70 70 90 00 98 45 13
68 | 50 73 70 90
69 | 36 27 00 04
70 | 05 40 22 11 51 50
71 | 31 69 68 05 65 13
72 | 09 80
```

The shortest and longest courses are 6433 and 7280 respectively. If the thousands' digits were used as the stem, there would be only two stems, 6 and 7. If they in turn were broken into 6l, 6h, 7l, 7h, there would be only four groups. Neither situation would yield a sufficient number of categories to reveal much information. If the first three leading digits were used as the stems, then there would be eighty-six stems, 643 to 728, and this would be too many.

Section 3.3

3.21 **a**

Number of Impairments	Frequency	Relative Frequency
0	100	.4167
1	43	.1792
2	36	.1500
3	17	.0708
4	24	.1000
5	9	.0375
6	11	.0458
	n = 240	1.0000

b   .4167 + .1792 + .1500 = .7459

c   1 − .7459 = .2541

d   .1000 + .0375 + .0458 = .1833

e   The frequencies (and relative frequencies) tend to decrease as the number of impairments increase.

3.23 **a**

Hits/Game	Frequency	Relative Frequency	Hits/Game	Frequency	Relative Frequency
0	20	0.0010	14	569	0.0294
1	72	0.0037	15	393	0.0203
2	209	0.0108	16	253	0.0131
3	527	0.0272	17	171	0.0088
4	1048	0.0541	18	97	0.0050
5	1457	0.0752	19	53	0.0027
6	1988	0.1026	20	31	0.0016
7	2256	0.1164	21	19	0.0010
8	2403	0.1240	22	13	0.0007
9	2256	0.1164	23	5	0.0003
10	1967	0.1015	24	1	0.0001
11	1509	0.0779	25	0	0.0000
12	1230	0.0635	26	1	0.0001
13	834	0.0430	27	1	0.0001
				19383	1.0005

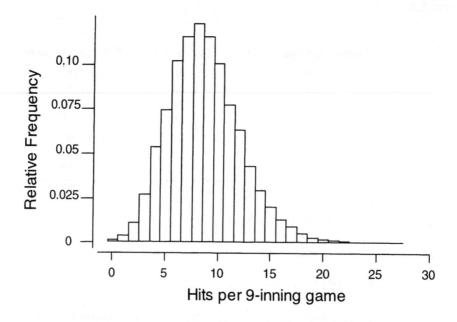

Hits per 9-inning game

b   The histogram rises rather smoothly to a single peak and then declines. The histogram stretches out a bit more on the right (toward large values) than it does on the left – a slight positive skew.

c   Proportion of games with at most 2 hits = proportion of games with 0 hits + proportion of games with 1 hit + proportion of games with 2 hits = .0010 + .0037 + .0108 = .0155

d   Proportion of games with between 5 and 10 hits = proportion of games with 6 hits + proportion of games with 7 hits + proportion of games with 8 hits + proportion of games with 9 hits = .1026 + .1164 + .1240 + .1164 = .4594

e   Proportion of games with more than 15 hits = .0131 + .0088 + .0050 + .0027 + .0016 + .001 + .0007 + .0003 + .0001 + .0000 + .0001 + .0001 = .0335

3.25   **a**

Concentration	Frequency	Rel. Freq.	Cum. Rel. Freq.
20 –< 30	1	.02	.02
30 –< 40	8	.16	.18
40 –< 50	8	.16	.34
50 –< 60	6	.12	.46
60 –< 70	16	.32	.78
70 –< 80	7	.14	.92
80 –< 90	2	.04	.96
90 –< 100	2	.04	1.00
	n = 50	1.00	

b   The proportion of the concentration observations that were less than 50 is .34. The proportion of the concentration observations that were at least 60 is $1 - .46 = .54$.

c   The proportion of the concentration observations in the interval $40 -< 70$ is $.78 - .18 = .60$.

d   No, for the 8 observations in the class $40 -< 50$, the frequency distribution does not contain any information as to how many are equal to 40.

3.27   a and b

Classes	Frequency	Rel. Freq.	Cum. Rel. Freq.
0 -< 6	2	.0225	.0225
6 -< 12	10	.1124	.1349
12 -< 18	21	.2360	.3709
18 -< 24	28	.3146	.6855
24 -< 30	22	.2472	.9327
30 -< 36	6	.0674	$1.0001 \approx 1.0$
	n = 50	1.0000	

c   (Rel. Freq. for $12 -< 18$) = (Cum. Rel. Freq. for < 18) − (Cum. Rel. Freq. for < 12)
$$= .3709 - .1349 = .2360.$$

d   The proportion that had pacemakers that did not malfunction within the first year equals 1 minus the proportion that had pacemakers that malfunctioned within the first year (12 months), which is $1 - .1349 = .8651$, which converts to 86.51%.

e   The proportion that required replacement between one and two years after implantation is equal to the proportion that had to be replaced within the first 2 years (24 months) minus the proportion that had to be replaced within the first year (12 months). This is $0.6855 - 0.1349 = 0.5506$, which converts to 55.06%.

f   The proportion that lasted less than 18 months is .3709, which converts to 37.09%, and the proportion that lasted less than 24 months is .6855, which converts to 68.55%. Thus, the time at which about 50% of the pacemakers failed is somewhere between 18 and 24 months. A more precise estimate can be found as follows.

$$\frac{(.50 - .3709)}{.3146} = \frac{x}{6} \Rightarrow x = \frac{6(.50 - .3709)}{.3146} = 2.46$$

So the time at which about 50% of the pacemakers had failed is $18 + 2.46$ or 20.46 months.

g
$$\frac{(.9237 - .9)}{.2472} = \frac{x}{6} \Rightarrow x = \frac{6(.9237 - .9)}{.2472} = .79$$

So an estimate of the time at which only 10% of the pacemakers initially implanted were still functioning is $30 - .79 = 29.21$ months.

3.29 **a**

Class Intervals	Frequency	Rel. Freq.	Density
.15 −< .25	8	.02192	0.2192
.25 −< .35	14	.03836	0.3836
.35 −< .45	28	.07671	0.7671
.45 −< .50	24	.06575	1.3150
.50 −< .55	39	.10685	2.1370
.55 −< .60	51	.13973	2.7946
.60 −< .65	106	.29041	5.8082
.65 −< .70	84	.23014	4.6028
.70 −< .75	11	.03014	0.6028
	n = 365	1.00001	

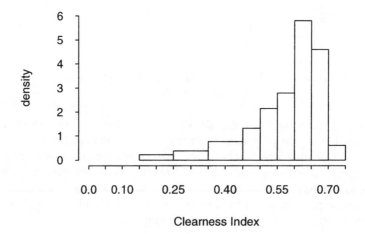

b The proportion of days with a clearness index smaller than .35 is $\dfrac{(8+14)}{365} = \dfrac{22}{365} = .06$, which converts to 6%.

c The proportion of days with a clearness index of at least .65 is $\dfrac{(84+11)}{365} = \dfrac{95}{365} = .26$, which converts to 26%.

3.31 Almost all the differences are positive indicating that the runners slow down. The graph is positively skewed. A typical difference value is about 150. About .02 of the runners ran the late distance more quickly that the early distance.

3.33   **a**

Number of cul-de-sacs	Frequency	Relative Frequency
0	17	0.3617
1	22	0.4681
2	6	0.1277
3	1	0.0213
4	0	0
5	1	0.0213
	47	1.0001

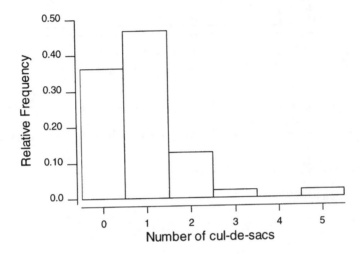

The proportion of subdivisions with no cul-de-sac is $\frac{17}{47}$ = .3617 or approximately

36%. The proportion of subdivisions with at least one cul-de-sac is $\frac{22+6+1+1}{47}$ =

.6383 or approximately 64%.

b

Number of intersections	Frequency	Relative Frequency
0	13	0.2766
1	11	0.2340
2	3	0.0638
3	7	0.1489
4	5	0.1064
5	3	0.0638
6	3	0.0638
7	0	0.0000
8	2	0.0426
	47	0.9999

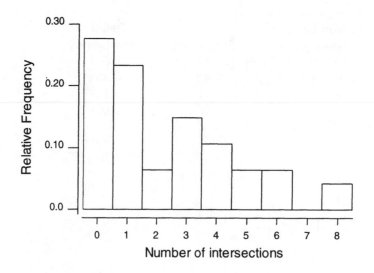

The proportion of subdivisions with at most 5 intersections =
$$\frac{13+11+3+7+5+3}{47} = \frac{42}{47} = .8936 \text{ or approximately 89\%. The proportion of}$$
subdivisions with fewer than 5 intersections is $\frac{13+11+3+7+5}{47} = \frac{39}{47} = .8298$ or
approximately 83%.

3.35   **a**

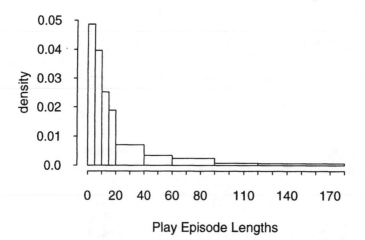

The density values corresponding to the nine class intervals are 0.0486, 0.0396, 0.0252, 0.0189, 0.0070, 0.0033, 0.0024, 0.0008 and 0.0006.

b   From the density histogram, the proportion of episodes lasting at least 20 seconds is approximately equal to

.0070(40 – 20) + .0033(60 – 40) + .0024(90 – 60) + .0008(120 – 90)
+ .0006(180 – 120) = .0070(20) + .0033(20) + .0024(30) + .0008(30) + .0006(60)
= .14 + .066 + .072 + .024 + .036 = .338.

The actual proportion is $\dfrac{(31+15+16+5+8)}{222} = \dfrac{75}{222} = .3378$.

c   From the density histogram, the proportion of episodes lasting between 40 and 75 seconds is approximately equal to
0.0033(60 – 40) + (.0024)(75 – 60) = .0033(20) + .0024(15) = .066 + .036 = .102.

From the frequency distribution, the proportion is approximately

$$\frac{\left(15+\dfrac{16}{2}\right)}{222} = \frac{23}{222} = .1036 \, .$$

3.37

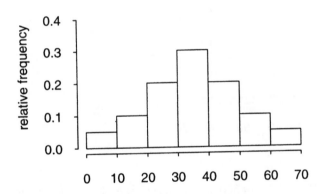

This histogram is symmetric.

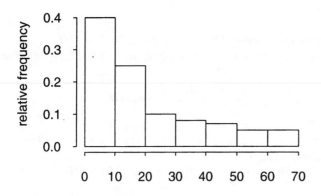

This histogram is positively skewed.

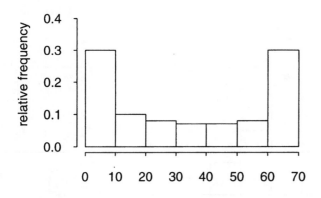

This is a bimodal histogram. While it is not perfectly symmetric it is close to being symmetric.

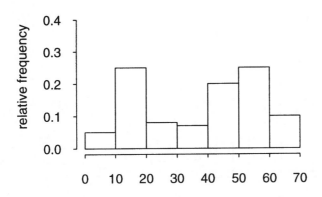

This is a bimodal histogram.

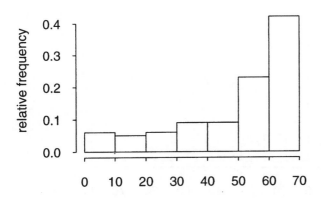

This is a negatively skewed histogram.

Supplementary Exercises

3.39   MINITAB represented a saturated fat content for margarine in the following way. A fat content of 17% would be represented with a stem of 1 and a leaf of 7. From the display it can be seen that the saturated fat content for these margarines ranged from a low of 9% to a high of 27%.

3.41

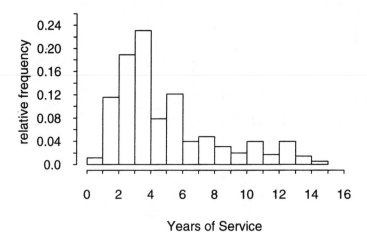

Years of Service

This histogram is positively skewed.

3.43   **a**

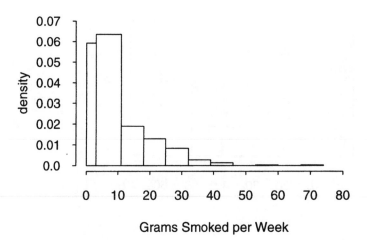

Grams Smoked per Week

b   The proportion of respondents who smoked 25 or more grams per week is $\dfrac{48}{529}$, or using the relative frequencies .0586 + .0189 + .0095 + .0019 + .0019 = .0908, which converts to 9.08%.

c   The approximate proportion that smoked between 15 and 18 grams per week is $\dfrac{(18-15)}{(18-11)}(.1323) = \dfrac{3}{7}(.1323) = .0567$. Thus, the approximate proportion that smoked more than 15 grams per week is 0.0567 + 0.0907 + 0.0586 + 0.0189 + 0.0095 + 0.0019 + 0.0019 = 0.2382, which converts to 23.82%.

d

Amount Smoked	Frequency	Rel. Freq.	Cumulative Rel. Frequency
0 −< 3	94	.1777	.1777
3 −< 11	269	.5085	.6862
11 −< 18	70	.1323	.8185
18 −< 25	48	.0907	.9092
25 −< 32	31	.0586	.9678
32 −< 39	10	.0189	.9867
39 −< 46	5	.0095	.9962
46 −< 53	0	.0000	.9962
53 −< 60	1	.0019	.9981
60 −< 67	0	.0000	.9981
67 −< 74	1	.0019	1.0000
	n = 529	1.0000	

The proportion of respondents who smoked at least 25 grams is 1 minus the proportion who smoked less than 25 grams. That is, $1 - .9092 = .0908$.

3.45   a

Class	Frequency	Rel. Freq.
.0 −< .1	0	.0000
.1 −< .2	0	.0000
.2 −< .3	0	.0000
.3 −< .4	6	.1304
.4 −< .5	13	.2826
.5 −< .6	14	.3043
.6 −< .7	5	.1087
.7 −< .8	3	.0652
.8 −< .9	3	.0652
.9 −< 1.0	0	.0000
1.0 −< 1.1	1	.0217
1.1 −< 1.2	1	.0217
	n = 46	.9998 ≈ 1.000

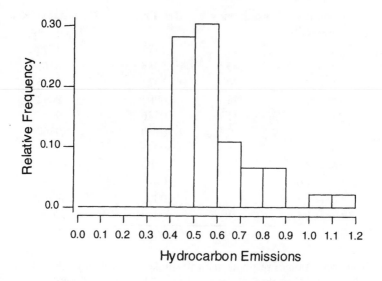

The HC values of 1.02 and 1.10 appear to be outliers.

b

Class	Frequency	Rel. Freq.
0 –< 2	1	.0217
2 –< 4	7	.1522
4 –< 6	15	.3261
6 –< 8	8	.1739
8 –< 10	3	.0652
10 –< 12	2	.0435
12 –< 14	2	.0435
14 –< 16	5	.1087
16 –< 18	0	.0000
18 –< 20	1	.0217
20 –< 22	0	.0000
22 –< 24	2	.0435
	n = 46	1.0000

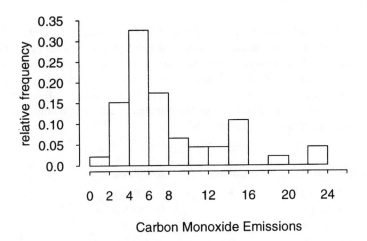

Carbon Monoxide Emissions

c   Both of the histograms are positively skewed.

3.47   **a**

```
1h │ 8 8 9 9 9
2l │ 0 0 0 0 1 1 1 1 2 2 3 4 4 4
2h │ 5 6 6 6 7 8
3l │ 0 0 1 2 2 4 4
3h │ 5 6 7 7 7              stem: hundreds
4l │ 0 1                    leaf: tens
4h │
5l │
5h │
6l │ 2
```

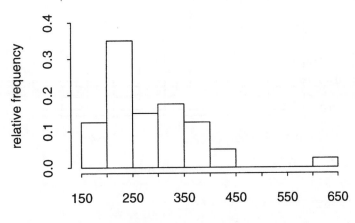

Length of novel (number of pages)

A typical book length is around 200 to 250 pages. Most of the books have a length
between 180 and 400. One book had 628 pages.

b   Answers will vary.

The only apparent similarity is that both distributions range from 0 to 300 minutes. The histogram for single-peak storms is skewed positively with a rather even decline as the minutes of duration increase. The duration of most single-peak storms were between 25 and 100 minutes. The histogram for multiple-peak storms is skewed negatively with a lot of variation in height of the rectangles. Most multiple-peak storms lasted either between 75 to 100 minutes or 225 to 300 minutes. This suggests that the distribution may be bimodal.

3.49 **a**

Class Intervals	Frequency	Rel. Freq.
0 −< 0.5	5	.1064
.5 −< 1.0	9	.1915
1.0 −< 1.5	10	.2128
1.5 −< 2.0	9	.1915
2.0 −< 2.5	3	.0638
2.5 −< 3.0	5	.1064
3.0 −< 3.5	1	.0213
3.5 −< 4.0	3	.0638
4.0 −< 4.5	1	.0213
4.5 −< 5.0	1	.0213
	n = 47	1.0001

b

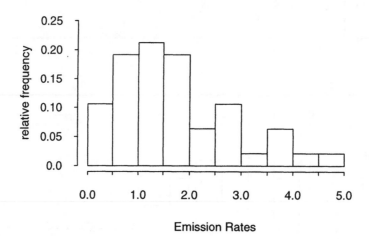

The histogram is positively skewed.

c

Class Intervals	Cumulative Rel. Freq.
0 –< 0.5	.1064
.5 –< 1.0	.2979
1.0 –< 1.5	.5107
1.5 –< 2.0	.7022
2.0 –< 2.5	.7660
2.5 –< 3.0	.8724
3.0 –< 3.5	.8937
3.5 –< 4.0	.9575
4.0 –< 4.5	.9788
4.5 –< 5.0	1.0001

d   i   About .2979 or 29.79% of the states had SO_2 emission below 1.0.

    ii   About .7022 – .2979 = .4043 or 40.43% of the states had SO_2 emission between 1.0 and 2.0.

    iii   About 1 – .7022 = .2978 or 29.78% of the states has SO_2 emission which exceeded 2.0.

3.51   a

Babies	Frequency	Relative Frequency
1	1	.0059
2	2	.0118
3	11	.0647
4	21	.1235
5	35	.2059
6	38	.2235
7	33	.1941
8	18	.1059
9	8	.0471
10	2	.0118
11	1	.0059
	170	1.0001

b   The proportion of the litters with more than 6 babies = $\dfrac{33+18+8+2+1}{170} = .3647$

The proportion with between 3 and 8 babies = $\dfrac{11+21+35+38+33+18}{170} = \dfrac{156}{170}$

= 0.9176

c   It is much easier to answer questions similar to those posed in (b) using the relative frequency table. The table gives the number of occurrences for each observation. This is the information needed to answer the questions. Using the raw data would require tallying for each question.

Chapter 4
Numerical Methods for Describing Data

Section 4.1

4.1 The data arranged in ascending order is:

7.6, 8.3, 9.3, 9.4, 9.4, 9.7, 10.4, 11.5, 11.9, 15.2, 16.2, 20.4

$$\overline{x} = \frac{139.3}{12} = 11.61$$

The sample median equals $\dfrac{9.7 + 10.4}{2} = 10.05$

The sample mean is somewhat larger than the sample median because of the outliers 15.2, 16.2 and 20.4. The sample median is more representative of a typical value since it is not influenced by the mentioned outliers.

4.3 The mean for the sample of 14 values is

$$\frac{13(119.7692) + 159}{14} = \frac{1557 + 159}{14} = \frac{1716}{14} = 122.57$$

4.5 **a**

32	55
33	49
34	
35	6699
36	34469
37	03345
38	9
39	2347
40	23
41	
42	4

stem : hundreds
leaf : ones

The stem and leaf display suggests that the mean and median will be fairly close to each other. Most values are between 356 and 375 and there are approximately equal amounts of values larger or smaller than these central values.

b $\bar{x} = \dfrac{9638}{26} = 370.69$

median $= \dfrac{369+370}{2} = 369.5$

c   The largest value, 424, could be increased by any arbitrary amount without affecting the sample median. The median is insensitive to outliers. However, if the largest value was decreased below the sample median, 369.5, then the value of the median would change.

4.7   **a**   Per capita personal income gives the average income per person in that area whereas total personal income gives the total income for the area. Areas are ranked by per capita personal income to give an idea of the average income for that area. Ranking by total personal income does not reflect the population size for that area.

b   West Palm Beach-Boca Raton must have a small population.

c   The approximate number of people in the San Jose area $= \dfrac{61,345,000,000}{37856} =$

1,620,483.

4.9   This statement could be correct if there were a small group of residents with very large wages. This group would make the average wage very large and thus a large percentage (perhaps even as large as 65%) could have wages less than the average wage.

4.11   The ordered values are:

Diet 1: 5 7 12 13 15
Diet 2: 3 5 10 11 13

Since each ordered observation in diet 1 is two units higher than the corresponding ordered observation in diet 2, it follows that the total for diet 1 is two units higher than the total for diet 2. Since the sample sizes are the same, it then follows that the average weight gain for diet 1 is two units higher than that for diet 2.

4.13   The sample median determines this salary. Its value is $\dfrac{4443+4129}{2} = \dfrac{8572}{2} = 4286.$

The mean salary paid in the six counties is

$$\dfrac{5354+5166+4443+4129+2500+2220}{6} = \dfrac{23812}{6} = 3968.67.$$

Since the mean salary is less than the median salary, the mean salary is less favorable to the supervisors.

4.15   The median and trimmed mean (trimming percentage of at least 20) can be calculated.

$$\text{sample median} = \frac{57+79}{2} = \frac{136}{2} = 68$$

$$20\% \text{ trimmed mean} = \frac{35+48+57+79+86+92}{6} = \frac{397}{6} = 66.17$$

4.17   The data arranged in ascending order is 4,8,11,12,33. The sample median is 11%.
The large outlier, 33%, has pulled the mean out towards that outlying value. The median, on the other hand, is not affected by the outlier and hence gives a better indication of a typical return.

Section 4.2

4.19 $n = 5$, $\Sigma x = 6$, $\bar{x} = \dfrac{6}{5} = 1.2$

x	$(x - \bar{x})$	$(x - \bar{x})^2$
1	−0.2	.04
0	−1.2	1.44
0	−1.2	1.44
3	1.8	3.24
2	0.8	0.64
6	0.0	6.80

$$s^2 = \frac{6.8}{4} = 1.70, \quad s = \sqrt{1.70} = 1.3038$$

4.21   **a**   Set 1: 2, 3, 7, 11, 12 ; $\bar{x} = 7$ and $s = 4.528$
Set 2: 5, 6, 7, 8, 9 ; $\bar{x} = 7$ and $s = 1.581$

   b   Set 1: 2, 3, 4, 5, 6 ; $\bar{x} = 4$ and $s = 1.581$
Set 2: 4, 5, 6, 7, 8 ; $\bar{x} = 6$ and $s = 1.581$

4.23 Subtracting 10 from each data point yields

x	$(x - \bar{x})$	$(x - \bar{x})^2$
52	13.636	185.9405
13	−25.364	643.3325
17	−21.364	456.4205
46	7.636	58.3085
42	3.636	13.2205
24	−14.364	206.3245
32	−6.364	40.5005
30	−8.364	69.9565
58	19.636	385.5725
35	−3.364	11.3165
73	34.636	1199.652

$$\bar{x} = \frac{422}{11} = 38.364$$

These deviations from the mean are identical to the deviations from the mean for the original data set. This would result in a variance for the new data set that is equal to the variance of the original data set. In general, adding the same number to each observation has no effect on the variance or standard deviation.

4.25

x	$x - \bar{x}$	$(x - \bar{x})^2$
244	51.4286	2644.9009
191	-1.5714	2.4693
160	-32.5714	1060.896
187	-5.5714	31.0405
180	-12.5714	158.0401
176	-16.5714	274.6113
174	-18.5714	344.8969
205	12.4286	154.4701
211	18.4286	339.6133
183	-9.5714	91.6117
211	18.4286	339.6133
180	-12.5714	158.0401
194	1.4286	2.0409
200	7.4286	55.1841
2696	.0004	5657.4285

$$\bar{x} = 192.5714$$

$$s^2 = \frac{5657.4285}{13} = 435.1868$$

$$s = 20.8611$$

s^2 could be interpreted as the mean squared deviation from the average leg power at a high workload. This is 435.1868. The standard deviation, s, could be interpreted as the typical amount by which leg power deviates from the average leg power. This is 20.8611.

4.27   lower quartile = 2.25 , upper quartile = 2.33,
iqr = 2.33 − 2.25 = .08,   iqr/1.35 = .08/1.35 = .059

The value of iqr/1.35 is less than s. This suggests that a histogram for this data would be heavy−tailed compared to a normal curve.

Section 4.3

4.29   **a**   The ordered data is:

33, 41, 41, 42, 44, 44, 46, 46, 46, 46, 48, 50, 50, 51, 53, 53, 54, 55, 60

The median value is 46.
The upper quartile is 53.
The lower quartile is 44.
The interquartile range is 53-44 = 9.

To check for mild outliers, we must compute 1.5 x 9 = 13.5. Are there any observations greater that 53 + 13.5 = 66.5?   No. Are there any observations less than 44 − 13.5 = 30.5? No. There are no mild outliers, therefore there are no extreme outliers either.

b

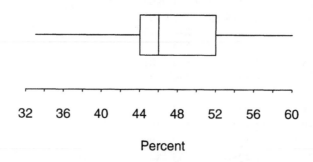

Percent

he median line is somewhat closer to the lower edge of the box suggesting a bit of positive skewness in the middle half of the data. The lower whisker is slightly longer than the upper whisker giving the impression of a slight negative skewness in the extremes of the data.

4.31   **a**   The median is $\dfrac{7.93+7.89}{2}=\dfrac{15.82}{2}=7.91$

The upper quartile is 8.01.
The lower quartile is 7.82
The interquartile range is 8.01 – 7.82 = 0.19.

b   To check for outliers, we calculate
1.5× iqr = 1.5× 0.19 = 0.285
3.0× iqr = 3.0× 0.19 = 0.57
The upper quartile + 1.5iqr = 8.01 + 0.285 = 8.295
The lower quartile – 1.5iqr = 7.82 – 0.285 = 7.535
The upper quartile + 3.0iqr = 8.01 + 0.57 = 8.58
The upper quartile – 3.0iqr = 7.82 – 0.57 = 7.25

There are no outliers in this data set.

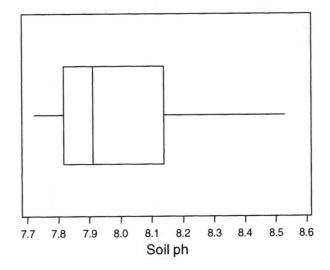

4.33   For Type 1, the ordered data is 655.5, 679.1, 699.4, 721.4, 734.3, 788.3.

The median = $\dfrac{699.4 + 721.4}{2}$ = 710.4
The lower quartile is 679.1.
The upper quartile is 734.3.
The interquartile range is 55.2

For Type 2, the ordered data is 686.1, 732.1, 772.5, 774.8, 786.9, 789.2.

The median = $\dfrac{772.5 + 774.8}{2}$ = 773.65
The lower quartile is 732.1.
The upper quartile is 786.9.
The interquartile range is 54.8.

For Type 3, the ordered data is 639.0, 671.2, 696.3, 717.2, 727.1, 737.1.

The median is $\dfrac{717.2+727.1}{2} = 722.15$

The lower quartile is 671.2
The upper quartile is 727.1.
The interquartile range is 55.9.

For Type 4, the ordered data is 520.0, 535.1, 542.4, 559.0, 586.9, 628.7.

The median is $\dfrac{542.4+559.0}{2} = 550.7$

The lower quartile is 535.1.
The upper quartile is 586.9.
The interquartile range is 51.8.

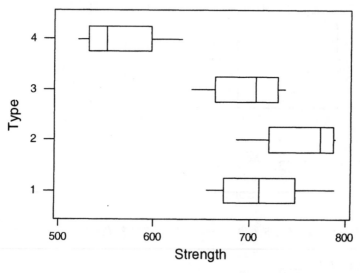

The median marker for Type 1 is centered in the boxplot whereas the medians for Type 2 and Type 3 are closer to the higher end of their boxplots and the median for Type 4 is closer to the lower end of its boxplot. The widths of the boxes are approximately equal indicating similar variability in the middle half of the data for each box type. There are no outliers for any of the different box types.

Section 4.4

4.35   **a**   The number $35 + 5 = 40$ is one standard deviation above the mean. The number $35 - 5 = 30$ is one standard deviation below the mean. The value 25 is 2 standard deviations below the mean and 45 is 2 standard deviations above the mean.

b   The value 25 is 2 standard deviations below the mean and 45 is 2 standard deviations above the mean. Thus, by Chebyshev's rule, at least 75% of the compact discs have playing times between 25 and 45 minutes.

c   The value 20 is 3 standard deviations below the mean and 50 is 3 standard deviations above the mean. Thus, by Chebyshev's rule, at least 89% of the playing times are between 20 and 50 minutes. This implies that no more than 11% of the compact discs have playing times which are either less than 20 minutes or greater than 50 minutes.

d   The value 25 is 2 standard deviations below the mean and 45 is 2 standard deviations above the mean. If the distribution of times is normal, then according to the empirical rule, roughly 95% of the compact discs will have playing times between 25 and 45 minutes. Approximately .3% will have playing times less than 20 minutes or greater than 50 minutes. Approximately .15% will have playing times of less than 20 minutes.

4.37 **a**   The value 59.60 is two standard deviations above the mean and the value 14.24 is two standard deviations below the mean. By Chebyshev's rule the percentage of observations between 14.24 and 59.60 is at least 75%.

b   The required interval extends from 3 standard deviations below the mean to 3 standard deviations above the mean. From $36.92 - 3(11.34) = 2.90$ to $36.92 + 3(11.34) = 70.94$.

c $\bar{x} - 2s = 24.76 - 2(17.20) = -9.64$

In order for the histogram for NO_2 concentration to resemble a normal curve, a rather large percentage of the readings would have to be less than 0 (since $\bar{x} - 25 = -9.64$). Clearly, a reading cannot be negative, hence the histogram cannot have a shape that resembles a normal curve.

4.39 For the first test the student's z-score is $\dfrac{(625 - 475)}{100} = 1.5$ and for the second test it is

$\dfrac{(45 - 30)}{8} = 1.875$. Since the student's z-score is larger for the second test than for the first test, the student's performance was better on the second exam.

4.41 Since the histogram is well approximated by a normal curve, the empirical rule will be used to obtain answers for part **a − c**.

a   Because 2500 is 1 standard deviation below the mean and 3500 is 1 standard deviation above the mean, about 68% of the sample observations are between 2500 and 3500.

b   Since both 2000 and 4000 are 2 standard deviations from the mean, approximately 95% of the observations are between 2000 and 4000. Therefore about 5% are outside the interval from 2000 to 4000.

c   Since 95% of the observations are between 2000 and 4000 and about 68% are between 2500 and 3500, there is about $95 - 68 = 27\%$ between 2000 and 2500 or 3500 and 4000. Half of those, $27/2 = 13.5\%$, would be in the region from 2000 to 2500.

d   When applied to a normal curve, Chebyshev's rule is quite conservative. That is, the percentages in various regions of the normal curve are quite a bit larger than the values given by Chebyshev's rule.

4.43   The recorded weight will be within 1/4 ounces of the true weight if the recorded weight is between 49.75 and 50.25 ounces. Now, $\dfrac{(50.25-49.5)}{.1} = 7.5$ and $\dfrac{(49.75-49.5)}{.1} = 2.5.$
Also, at least $1 - 1/(2.5)^2 = 84\%$ of the time the recorded weight will be between 49.25 and 49.75. This means that the recorded weight will exceed 49.75 no more than 16% of the time. This implies that the recorded weight will be between 49.75 and 50.25 no more than 16% of the time. That is, the proportion of the time that the scale showed a weight that was within 1/4 ounce of the true weight of 50 ounces is no more than 0.16.

4.45   Because the number of answers changed from right to wrong cannot be negative and because the mean is 1.4 and the value of the standard deviation is 1.5, which is larger than the mean, this implies that the distribution is skewed positively and is not a normal curve. Since $(6 - 1.4)/1.5 = 3.07$, by Chebyshev's rule, at most $1/(3.07)^2 = 10.6\%$ of those taking the test changed at least 6 from correct to incorrect.

4.47   **a**

Class	Frequency	Rel Freq.	Cum. Rel. Freq.
5 –< 10	13	.26	.26
10 –< 15	19	.38	.64
15 –< 20	12	.24	.88
20 –< 25	5	.10	.98
25 –< 30	1	.02	1.00
	n = 50	1.00	

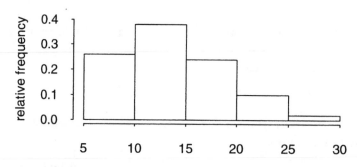

1989 per capita expenditures on libraries

b   **i**   The 50th percentile is between 10 and 15 since the cumulative relative frequency at 10 is .26 and at 15 it is .64. The 50th percentile is approximately $10 + 5(50 - 26)/38 = 10 + 3.158 = 13.158.$

ii The 70th percentile is between 15 and 20 since the cumulative relative frequency at 15 is .64 and at 20 it is .88. The 70th percentile is approximately $15 + 5(70 - 64)/24 = 15 + 1.25 = 16.25$.

iii The 10th percentile is between 5 and 10 and is approximately $5 + 5(10 - 0)/26 = 5 + 1.923 = 6.923$.

iv The 90th percentile is between 20 and 25 and is approximately $20 + 5(90 - 88)/10 = 20 + 1 = 21$.

v The 40th percentile is between 10 and 15 and is approximately $10 + 5(40 - 26)/38 = 10 + 1.842 = 11.842$.

Supplementary Exercises

4.49 **a**

Cancer		No Cancer
9 6 8 3 7 9 5	0	9 5 7 6 8 3 9 7 6 7 8 9 9 3
8 6 0 7 1 8 1 5 0 6 6 8 1 5 2 3 3 1 5 0	1	1 2 2 7 1 7 1 3 1 1 4
1 2 3 0 2 7 3 1	2	9 9 4 9 4 1 9 1
8 3 4 9	3	8 3 9
5	4	
7	5	5 5
	6	
	7	
HI: 210	8	5

Both data sets are positively skewed, but the displays differ in the first two stems.

b For the cancer group: $\bar{x} = \dfrac{958}{42} = 22.81$ and the median $= \dfrac{(16+16)}{2} = 16$.

For the no-cancer group: $\bar{x} = \dfrac{747}{39} = 19.15$ and the median $= 12$.

The values of both the mean and median suggest that the center of the cancer sample is to the right (larger in value) of the center of the no-cancer group.

c For the cancer group:

$$s^2 = \frac{S_{xx}}{n-1} = \frac{41084.476}{41} = 1002.0604$$

$$s = \sqrt{1002.0604} = 31.6553$$

For the no-cancer group:

$$s^2 = \frac{S_{xx}}{n-1} = \frac{10969.077}{38} = 288.6599$$

$$s = \sqrt{288.6599} = 16.99$$

The values of s^2 and s suggest that there is more variability in the cancer sample values than in the no-cancer sample values.

d   For the cancer group, the iqr = 22 – 11 = 11.
For the no-cancer group, the iqr = 24 – 8 = 16.
The previous conclusion about variability is not confirmed by the interquartile ranges. A possible explanation is the extreme outlier in the cancer sample whose value is 210. Eliminating the largest value in each group and recalculating the variances yields the following results.

For the cancer group: $n = 41$, $\sum x = 748$, $\sum x^2 = 18836$

$$s^2 = \frac{18836 - \frac{(748)^2}{41}}{40} = \frac{18836 - 13646.439}{40} = \frac{5189.561}{40} = 129.739$$

$$s = \sqrt{129.739} = 11.39$$

For the no-cancer group: $n = 38$, $\sum x = 662$, $\sum x^2 = 18052$

$$s^2 = \frac{18052 - \frac{(662)^2}{38}}{37} = \frac{18052 - 11532.7368}{37} = \frac{6519.2632}{37} = 176.1963$$

$$s = \sqrt{176.1963} = 13.27$$

The values of s for the two groups with the outliers removed are closer in value and the no-cancer group is higher.

e

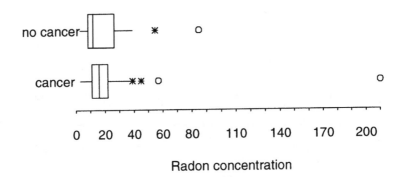

Radon concentration

4.51 It implies that the median salary is less than $1 million and that the histogram would have a shape which is positively skewed.

4.53 **a**

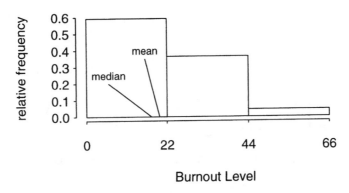

Burnout Level

b The two measures of center are not identical because the histogram is not symmetric.

c Using Chebyshev's rule, we are guaranteed to find at least 75% of the scores in the interval extending from
$$\bar{x} - 2s = 19.93 - 2(12.89) = -5.85 \; to \; \bar{x} + 2s = 19.93 + 2(12.89) = 45.71.$$

Since a burnout level score cannot be negative, we could change the left-hand endpoint of −5.85 to 0. Therefore, we are guaranteed to find at least 75% of the scores in the interval extending from 0 to 45.71. From the histogram, about (554+342)/937 = .956, or roughly 96% fall in this interval.

4.55

	Smoking	Nonsmoking
$\bar{x}$	693.364	961.091
Median	693.000	947.000
lower quartile	598.000	903.000
upper quartile	767.000	981.000
S^2	10787.700	12952.700
S	103.864	113.810

Boxplots for the two groups are below.

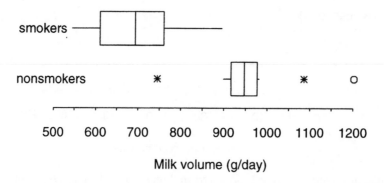

The data strongly suggests that milk volume for nonsmoking mothers is greater than that for smoking mothers.

4.57   **a**

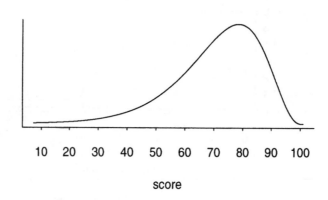

b

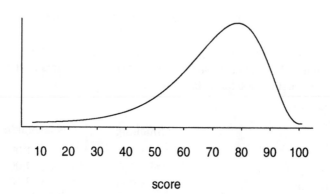

4.59   Since the mean is larger than the median, this suggests that the distribution of values is positively skewed or has some outliers with very large values.

4.61

x	$(x - \bar{x})$	$(x - \bar{x})^2$
18	−4.15	17.22
18	−4.15	17.22
25	2.85	8.12
19	−3.15	9.92
23	0.85	0.72
20	−2.15	4.62
69	46.85	2194.92
18	−4.15	17.22
21	−1.15	1.32
18	−4.15	17.22
18	−4.15	17.22
20	−2.15	4.62
18	−4.15	17.22
18	−4.15	17.22
20	−2.15	4.62
18	−4.15	17.22
19	−3.15	9.92
28	5.85	34.22
17	−5.15	26.52
18	−4.15	17.22
443	0.00	2454.48

a $\bar{x} = \dfrac{443}{20} = 22.15$

$s^2 = 2454.48/19 = 129.183$ and $s = \sqrt{129.183} = 11.366$

b   The 10% trimmed mean is calculated by eliminating the two largest values (69 and 28) and the two smallest values (17 and 18). The trimmed mean equals 311/16 = 19.4375. It is a better measure of location for this data set since it eliminates a very large value (69) from the calculation. It is almost 3 units smaller than the average.

c   The upper quartile is (21 + 20)/2 = 20.5, the lower quartile is (18 + 18)/2 = 18, and the iqr = 20.5 − 18 = 2.5.

d   upper quartile + 1.5(iqr) = 20.5 + 1.5(2.5) = 24.25
upper quartile + 3.0(iqr) = 20.5 + 3(2.5) = 28.00
The values 25 and 28 are mild outliers and 69 is an extreme outlier.

Chapter 5
Summarizing Bivariate Data

Section 5.1

5.1   **a**

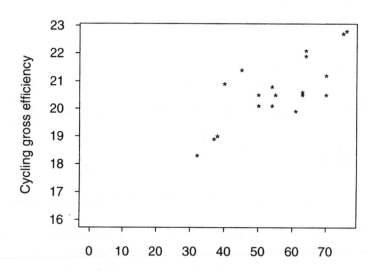

Percentage of type I (slow twitch) muscle fibers

There is a tendency for y to increase as x does. That is, larger values of gross efficiency tend to be associated with larger values of percentage type I fibers (a positive relationship between the variables).

      b   There are several observations that have identical x-values yet different y-values (for example, $x_6 = x_7 = 50$, but $y_6 = 20.5$ and $y_7 = 20.1$). Thus, the value of y is *not* determined solely by x, but also by various other factors.

5.3   **a**   For x = BOD mass, the median is $\dfrac{27+30}{2} = \dfrac{57}{2} = 28.5$.

The lower quartile is 11 and the upper quartile is 38.
The iqr is 38 − 11 = 27.

To check for outliers, we compute
$1.5 \times iqr = 1.5 \times 27 = 40.5$ and $3.0 \times iqr = 3.0 \times 27 = 81$.
Therefore, 103 is a mild outlier and 142 is an extreme outlier.

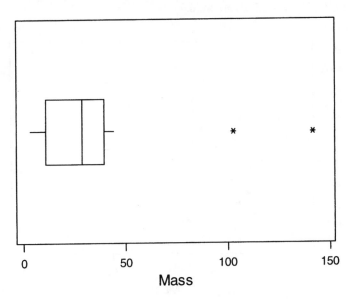

Mass

For y = BOD removal, the median is $\dfrac{11+16}{2} = \dfrac{27}{2} = 13.5$.

The lower quartile is 8 and the upper quartile is 30.
The iqr is $30 - 8 = 22$.
To check for outliers, we compute
$1.5 \times iqr = 1.5 \times 22 = 33.0$ and $3.0 \times iqr = 3.0 \times 22 = 66$.
Therefore, 75 and 90 are outliers.

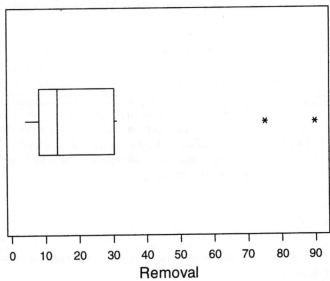

Removal

The medians are not centered for either of the boxplots suggesting skewness in the middle half of the data. There are 2 outliers for the BOD mass boxplot, one mild, 103 and one extreme, 142. There are also 2 outliers for the BOD removal boxplot, 75 and 90. Both of these outliers are mild.

b

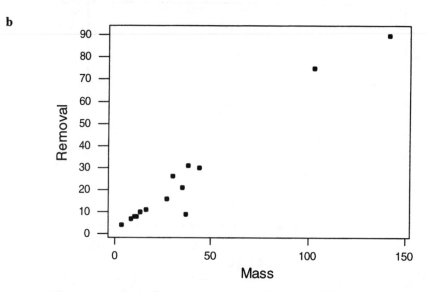

There is a tendency for y to increase as x increases. That is, the larger values for BOD mass will be associated with larger values for BOD removal.

5.5   There are several observations that have identical or nearly identical x-values yet different y-values. Therefore, the value of y is not determined solely by x, but also by various other factors. There appears to be a general tendency for y to decrease in value as x increases in value. There are two data points which are far removed from the remaining data points. These two data points have large x-values and small y-values. Their presence might have an undue influence on a line fit to the data.

5.7   **a**   There are several values that have identical or nearly identical x-values yet different y-values. Therefore, the value of y is not determined solely by x , but also by various other factors. There appears to be a tendency for y to decrease as x increases.

       b   People with low body weight tend to be small people and it is possible their livers may be smaller than the liver of an average person. Conversely, people with high weight tend to be large people and their livers may be larger than the liver of an average person. Therefore, we would expect the graft weight ratio to be large for low weight people and small for high weight people.

Section 5.2

5.9   **a**   A positive correlation would be expected, since as temperature increases cooling costs would also increase.

b　A negative correlation would be expected, since as interest rates climb fewer people would be submitting applications for loans.

c　A positive correlation would be expected, since husbands and wives tend to have jobs in similar or related classifications. That is, a spouse would be reluctant to take a low-paying job if the other spouse had a high-paying job.

d　No correlation would be expected, because those people with a particular I.Q. level would have heights ranging from short to tall.

e　A positive correlation would be expected, since people who are taller tend to have larger feet and people who are shorter tend to have smaller feet.

f　A weak to moderate positive correlation would be expected. There are some who do well on both, some who do poorly on both, and some who do well on one but not the other. It is perhaps the case that those who score similarly on both tests outnumber those who don't.

g　A negative correlation would be expected, since there is a fixed amount of time and as time spent on homework increases, time in watching television would decrease.

h　No correlation overall, because for small or substantially large amounts of fertilizer yield would be small.

5.11

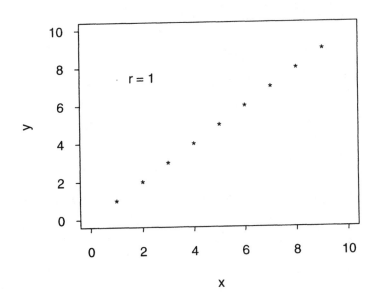

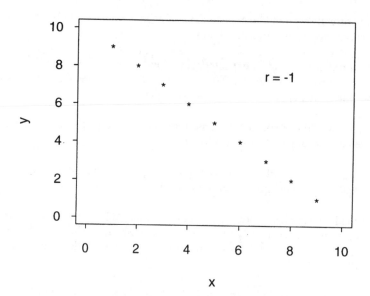

5.13 **a** Since the points tend to be close to the line, it appears that x and y are strongly correlated in a positive way.

       b An r value of .9366 indicates a strong positive linear relationship between x and y.

       c If x and y were perfectly correlated with r = 1, then each point would lie exactly on a line. The line would not necessarily have slope 1 and intercept 0.

5.15 $r = \dfrac{27918 - \dfrac{(9620)(7436)}{2600}}{\sqrt{36168 - \dfrac{(9620)(9620)}{2600}}\sqrt{23145 - \dfrac{(7436)(7436)}{2600}}}$

$= \dfrac{27918 - 27513.2}{\sqrt{574}\sqrt{1878.04}} = \dfrac{404.8}{(23.9583)(43.336)} = .3899$

5.17 No. An r value of −0.085 indicates an extremely weak relationship between support for environmental spending and degree of belief in God.

5.19 The sign of r is determined by the numerator quantity $\sum xy - \dfrac{\sum x \sum y}{n}$. Let w be the missing y-value. From the data,

n = 5, $\sum x = 15$, $\sum y = 10 + w$, $\sum xy = 30 + 5w$ and so

$\sum xy - \dfrac{\sum x \sum y}{n} = (30 + 5w) - \dfrac{15(10 + w)}{5} = 30 + 5w - 30 - 3w = 2w$

Since $w \geq 0, \sum xy - \dfrac{\sum x \sum y}{n} = 2w \geq 0$ and hence r cannot be negative.

5.21

x	y	$\sum xy$
3	5	15
5	3	15
1	11	11
7	6	42
2	7	14
11	4	44
9	10	90
12	2	24
8	9	72
10	1	10
4	12	48
15	13	195
14	8	112
6	14	84
13	15	195
		971

$$r_s = \dfrac{971 - \dfrac{(15)(16)(16)}{4}}{\dfrac{(15)(14)(16)}{12}} = \dfrac{971 - 960}{280} = \dfrac{11}{280} = 0.039$$

A value of r = 0.039 indicates a weak relationship between frequency of sex and partners reported satisfaction.

5.23 $\sum (x \text{ rank})(y \text{ rank}) = 4843$

$$r_s = \dfrac{4843 - \dfrac{24(25)^2}{4}}{\dfrac{24(23)(25)}{12}} = \dfrac{1093}{1150} = .9504$$

This indicates a very strong positive relationship between the 1993 ranks and the 1994 ranks.

5.25 **a**

d:	0	1	0	−1	0	1	−1	−1	1
d²:	0	1	0	1	0	1	1	1	1

$\sum d^2 = 6$

$$r_s = 1 - \frac{6(6)}{9(80)} = 1 - \frac{36}{270} = 1 - .05 = .95$$

b   When there is a perfect positive relationship, each d will be zero because the x and y ranks for an observation will be the same. Thus $\sum d^2 = 0$ and r_s will equal 1.

Section 5.3

5.27   **a**   For this data set, n= 13, $\sum x = 5.92$, $\sum x^2 = 3.8114$, $\sum y = 10.47$,

$$\sum y^2 = 9.885699, \quad \sum xy = 5.8464, \quad \bar{x} = 0.455, \quad \bar{y} = 0.805$$

$$b = \frac{5.8464 - \dfrac{(5.92)(10.47)}{13}}{3.8114 - \dfrac{(5.92)(5.92)}{13}} = \frac{5.8464 - 4.7679}{3.8114 - 2.6959} = \frac{1.0785}{1.1155} = 0.9668$$

$$a = 0.805 - 0.9668(0.455) = 0.3651$$

The least squares regression line is $\hat{y} = 0.3651 + 0.9668x$

b   For a value of x = 0.5, $\hat{y} = 0.3651 + 0.9668(0.5) = 0.8485$

5.29   **a**   slope = 244.9   intercept = −275.1

        b   244.9

        c $y = -275.1 + 244.9(2) = -275.1 + 489.8 = 214.7$

        d   No. When shell height (x) equals 1, the equation would result in a predicted breaking strength of $-275.1 + 244.9(1) = -30.2$. It is impossible for breaking strength to be a negative value, so the equation results in a predicted value which is not meaningful.

5.31   **a** $\hat{y} = 31.040 - 5.792x$

        b   −5.792; i.e., a decrease of 5.792 in percent crown dieback.

        c $\hat{y} = 31.040 - 5.792(4.0) = 7.872$

        d   When $x = 5.6$, $\hat{y} = 31.040 - 5.792(5.6) = -1.3952$. Since mean crown dieback cannot be negative, it would not be sensible to use the least squares line to predict percent dieback when soil pH has a value of 5.6. Also, 5.6 is outside the range of x values used to calculate the equation coefficients.

5.33   It is dangerous to use the least squares line to obtain predictions for x-values outside the range of those contained in the sample, because there is no information in the sample about the relationship that exists between x and y beyond the range of the data. The relationship may be the same or it may change substantially. There is no data to support a conclusion either way.

5.35   The denominators of b and of r are always positive numbers. The numerator of b and r is $\sum(x-\bar{x})(y-\bar{y})$. Since both b and r have the same numerator and positive denominators, they will always have the same sign.

Section 5.4

5.37   **a**   A value of $r^2 = 0.7664$ means that 76.64% of the observed variability in clutch size can be explained by an approximate linear relationship between clutch size and snout vent length.

b   To find the value of s_e, we need the value of SSResid.

We know $r^2 = 1-$ SSResid/SSTo

Solving for SSResid, we get, SSResid = 10266.954

$$s_e = \sqrt{\frac{10266.954}{14-2}} = 29.25$$

Thus, a typical amount by which an observation deviates from the least squares line is 29.25.

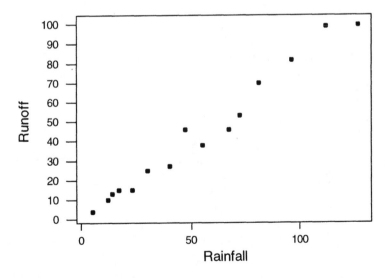

5.39   **a** $\hat{y} = 766 + .015(9900) = 914.5$

residual $= 893 - 914.5 = -21.5$

b   The typical amount that average SAT score deviates from the least squares line is 53.7.

c   Only about 16% of the observed variation in average SAT scores can be attributed to the approximate linear relationship between average SAT scores and expenditure per pupil. The least-squares line does not effectively summarize the relationship between average SAT scores and expenditure per pupil.

5.41   **a**   The least square line is $\hat{y} = 32.08 + 0.5549x$

x	y	predicted	residual
15	23	40.4048	−17.4048
19	52	42.6245	9.3755
31	65	49.2837	15.7163
39	55	53.7231	1.2769
41	32	54.8330	−22.8330
44	60	56.4978	3.5022
47	78	58.1626	19.8374
48	59	58.7175	0.2825
55	61	62.6020	−1.6020
65	60	68.1513	−8.1513

b $\text{SSResid} = (-17.4048)^2 + (9.3755)^2 + \ldots + (-1.6020)^2 + (-8.1513)^2$
$$= 302.9271 + 87.9000 + \ldots + 2.5664 + 66.4437 = 1635.6833$$

$$\text{SSTo} = 31993 - \frac{(545)^2}{10} = 31993 - 29702.5 = 2290.50$$

$$r^2 = 1 - \frac{1635.6833}{2290.5000} = 1 - .7141 = .2859$$

c   The least squares line does not give very accurate predictions. Only 28.59% of the observed variation in age is explained by the linear relationship between percent of root transparent dentine and age.

5.43   **a** $\sum x^2 - \dfrac{(\sum x)^2}{n} = 62.600235 - \dfrac{(22.027)^2}{12} = 62.600235 - 40.43239 = 22.16784$

$\sum xy - \dfrac{(\sum x \sum y)}{n} = 1114.5 - \dfrac{(22.027)(793)}{12} = 1114.5 - 1455.61758 = -341.11758$

$b = \dfrac{-341.11758}{22.16784} = -15.38795$

$a = 66.08333 - (-15.38795)(1.83558) = 66.08333 + 28.24586 = 94.32919$

The least squares equation is $\hat{y} = 94.33 - 15.388x$

b $$\text{SSTo} = \sum y^2 - \frac{(\sum y)^2}{n} = 57939 - \frac{(793)^2}{12} = 57939 - 52404.08333 = 5534.91667$$

$$\text{SSResid} = 57939 - 94.32919(793) - (-15.38795)(1114.5)$$
$$= 57939 - 74803.04767 + 17149.87028 = 285.82261$$

c $$r^2 = 1 - \frac{285.82261}{5534.91667} = 1 - .05164 = .94836 \text{ or } 94.836\%$$

d $$s_e^2 = \frac{285.82261}{10} = 28.582261$$

$$s_e = \sqrt{28.582261} = 5.34624$$

e Since the slope of the fitted line is negative, the value of r is the negative square root of r^2. So $r = -\sqrt{r^2} = -\sqrt{.94836} = -.97384$.

5.45 $$r^2 = 1 - \frac{1235.470}{25321.368} = 0.9512$$

The coefficient of determination reveals that 95.12% of the total variation in hardness of molded plastic can be explained by the linear relationship between hardness and the amount of time elapsed since termination of the molding process.

5.47 From problem **5.43**, the equation of the least-squares line is $\hat{y} = 94.33 - 15.388x$.

x	y	$\hat{y}$	residual
.106	98	92.6989	5.30112
.193	95	91.3601	3.63988
.511	87	86.4667	0.53326
.527	85	86.2205	-1.22053
1.08	75	77.7110	-2.71096
1.62	72	69.4014	2.59856
1.73	64	67.7088	-3.70876
2.36	55	58.0143	-3.01432
2.72	44	52.4746	-8.47464
3.12	41	46.3194	-5.31945
3.88	37	34.6246	2.37544
4.18	40	30.0082	9.99184

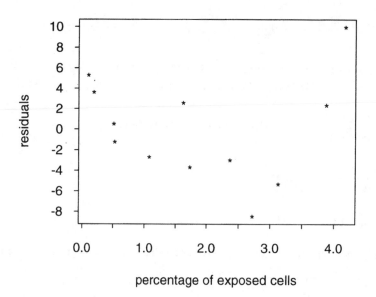

percentage of exposed cells

There appears to be a pattern in the plot. It is like the graph of a quadratic equation.

5.49 **a** Whether s_e is small or not depends upon the physical setting of the problem. An s_e of 2 feet when measuring heights of people would be intolerable, while an s_e of 2 feet when measuring distances between planets would be very satisfactory. It is possible for the linear association between x and y to be such that r^2 is large and yet have a value of s_e that would be considered large. Consider the following two data sets:

Set 1		Set 2	
x	y	x	y
5	14	14	5
6	16	16	15
7	17	17	25
8	18	18	35
9	19	19	45
10	21	21	55

For set 1, $r^2 = .981$ and $s_e = .378$. For set 2, $r^2 = .981$ and $s_e = 2.911$.
Both sets have a large value for r^2, but s_e for data set 2 is 7.7 times larger than s_e for data set 1. Hence, it can be argued that data set 2 has a large r^2 and a large s_e.

b Now consider the data set

x	5	55	15	45	25	35
y	10.004	10.006	10.007	10.008	10.009	10.010

This data set has $r^2 = .12$ and $s_e = .002266$. So yes, it is possible for a bivariate data set to have both r^2 and s_e small.

c   When r^2 is large and s_e is small, then not only has a large proportion of the total variability in y been explained by the linear association between x and y, but the typical error of prediction is small.

5.51   **a**   When $r = 0$, then $s_e = s_y$. The least squares line in this case is a horizontal line with intercept of $\overline{y}$.

b   When r is close to 1 in absolute value, then s_e will be much smaller than s_y.

c $s_e = \sqrt{1-(.8)^2}(2.5) = .6(2.5) = 1.5$

d   Letting y denote height at age 6 and x height at age 18, the equation for the least squares line for predicting height at age 6 from height at age 18 is

$$\text{(height at age 6)} = 46 + .8\left(\frac{1.7}{2.5}\right)[\text{(height at age 18)} - 70].$$

The value of s_e is $\sqrt{1-(.8)^2}\,(1.7) = .6(1.7) = 1.02$.

Section 5.5

5.53   **a**

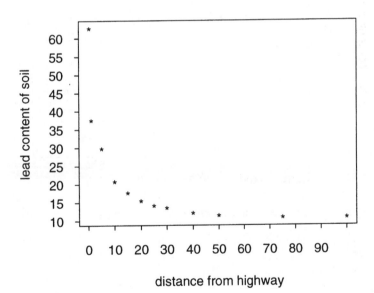

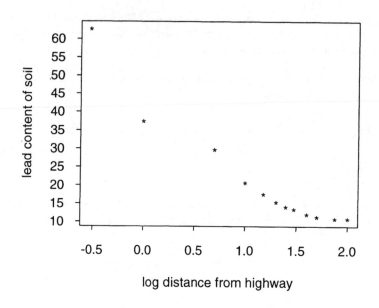

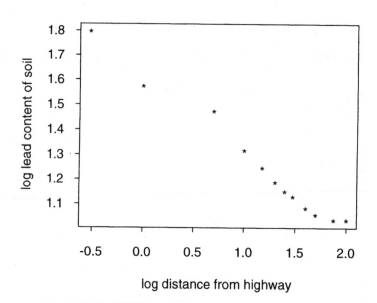

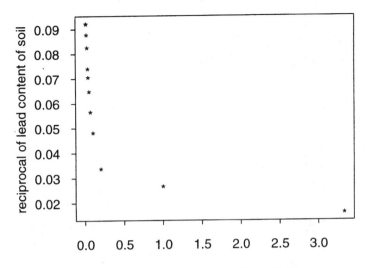

reciprocal of distance from highway

b   It appears that log x, log y does the best job of producing an approximate linear relationship. The least-squares equation for predicting $y' = \log y$ from $x' = \log x$ is

$$\hat{y}' = 1.61867 - .31646x'.$$

When $x = 25, x' = 1.39764$, $\hat{y}' = 1.61867 - .31646(1.39764) = 1.17628$

$$\hat{y} = 10^{1.17628} = 15.0064$$

5.55   **a** $n = 12$ $\sum x = 22.4$ $\sum y = 303.1$ $\sum x^2 = 88.58$ $\sum y^2 = 12039.27$

$\sum xy = 241.29$

$$r = \frac{241.29 - \dfrac{(22.4)(303.1)}{12}}{\sqrt{88.58 - \dfrac{(22.4)^2}{12}}\sqrt{12039.27 - \dfrac{(303.1)^2}{12}}} = \frac{-324.50}{\sqrt{46.767}\sqrt{4383.47}}$$

$$= \frac{-324.5}{(6.84)(66.208)} = -0.717$$

b $n = 12$ $\sum x = 13.5$ $\sum y = 55.74$ $\sum x^2 = 22.441$ $\sum y^2 = 303.3626$

$\sum xy = 47.7283$

$$r = \cfrac{47.7283 - \cfrac{(13.5)(55.74)}{12}}{\sqrt{22.441 - \cfrac{(13.5)^2}{12}}\sqrt{303.3626 - \cfrac{(55.74)^2}{12}}} = \cfrac{-14.9792}{\sqrt{7.2535}\sqrt{44.4503}}$$

$$\cfrac{-14.9792}{(2.693)(6.667)} = -0.835$$

The correlation between $\sqrt{x}$ and $\sqrt{y}$ is $-.835$. Since this correlation is larger in absolute value than the correlation of part **a**, the transformation appears successful in straightening the plot.

5.57   **a**

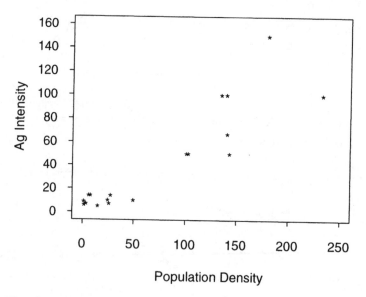

The plot does appear to have a positive slope, so the scatter plot is compatible with the "positive association" statement made in the paper.

b

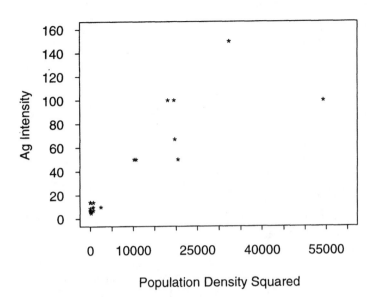

This transformation does straighten the plot, but it also appears that the variability of y increases as x increases.

c

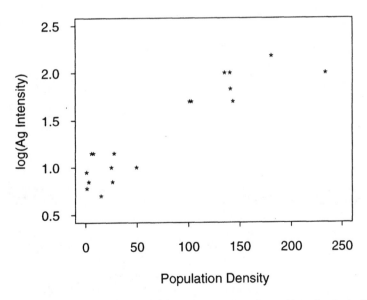

The plot appears to be as straight as the plot in **b**, and has the desirable property that the variability in y appears to be constant regardless of the value of x.

d

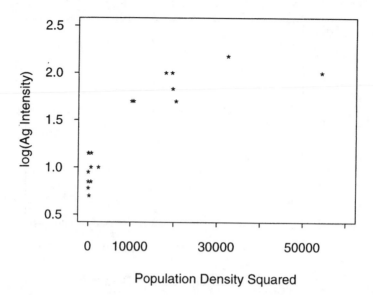

Population Density Squared

This plot has curvature opposite of the plot in **a**, suggesting that this transformation has taken us too far along the ladder.

5.59

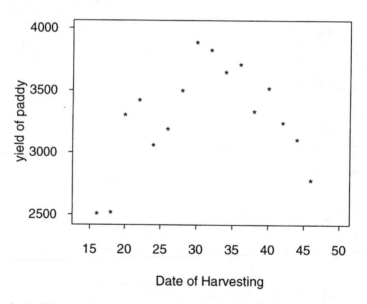

Date of Harvesting

This plot is not like any of the four segments of Figure 5.34. This suggests that transforming x or y by the methodology of this section might not be productive in straightening the plot. The plot resembles the graph of a quadratic equation, and hence, a parabola might provide a reasonable fit to the plot.

Supplementary Exercises

5.61 **a** The plot does not suggest a linear relationship. However, the one outlier value (51.3, 49.3) prevents an accurate interpretation.

b $\hat{y} = -11.37 + 1.0906(40) = 32.254$

c About 59.5 percent of the variability in fire-simulation consumption is explained by the linear relationship between treadmill consumption and fire-simulation consumption.

d For the new data set, $n = 9$, $\Sigma x = 388.8 - 51.3 = 337.5$, $\Sigma y = 310.3 - 49.3 = 261.0$

$\Sigma x^2 = 15338.54 - (51.3)^2 = 12706.85$, $\Sigma y^2 = 10072.41 - (49.3)^2 = 7641.92$

$\Sigma xy = 12306.58 - (51.3)(49.3) = 9777.49$

$$\Sigma x^2 - \frac{(\Sigma x)^2}{n} = 12706.85 - \frac{(337.5)^2}{9} = 50.60$$

$$\Sigma xy - \frac{(\Sigma x)(\Sigma y)}{n} = 9777.49 - \frac{(337.5)(261.0)}{9} = -10.01$$

$$b = \frac{-10.01}{50.60} = -0.1978, \quad a = \frac{261}{9} - (-.1978)\left(\frac{337.5}{9}\right) = 36.4175$$

$\hat{y} = 36.4175 - .1978x$

$$\Sigma y^2 - \frac{(\Sigma y)^2}{n} = 7641.92 - \frac{(261)^2}{9} = 72.92, \quad r^2 = \frac{(-10.01)^2}{(50.6)(72.92)} = .027$$

Without the observation (51.3, 49.3) there is very little evidence of a linear relationship between fire-simulation consumption and treadmill consumption. One would be very hesitant to use the prediction equation based on the data set including this observation because this observation is very influential.

5.63 The summary values are: $n = 13$, $\Sigma x = 91$, $\Sigma y = 470$, $\Sigma x^2 = 819$, $\Sigma y^2 = 19118$

$\Sigma xy = 3867$

$\Sigma xy - \frac{(\Sigma x)(\Sigma y)}{n} = 577$, $\Sigma x^2 - \frac{(\Sigma x)^2}{n} = 182$, $\Sigma y^2 - \frac{(\Sigma y)^2}{n} = 2125.6923$

a $b = \frac{577}{182} = 3.1703$ $a = 36.1538 - 3.1703(7) = 13.9617$

The equation of the estimated regression line is $\hat{y} = 13.9617 + 3.1703x$

b

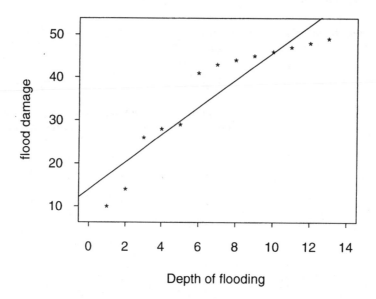

Depth of flooding

The plot with the line drawn in suggests that perhaps a simple linear regression model may not be appropriate. The scatterplot suggests that a curvilinear relationship may exist between flood depth and damage. The points for small x-values or large x-values are below the line, while points for x-values in the middle range are above the line.

c   When x = 6.5, $\hat{y}$ = 13.9617 + 3.1703(6.5) = 34.5687

d   The scatterplot in **b** suggests that the value of damage levels off at between 45 and 50 when the depth of flooding is in excess of 10 feet. Using the least squares line to predict flood damage when x = 18 would yield a very high value for damage and result in a predicted value far in excess of actual damage. Since x = 18 is outside of the range of x-values for which data has been collected, we have no information concerning the relationship in the vicinity of x = 18. All of these reasons suggest that one would not want to use the least squares line to predict flood damage when depth of flooding is 18 feet.

The relationship between the two r^2's is that they are equal in value.

5.65 **a** $\sum(x-\bar{x})^2 = 2,\ \sum(y-\bar{y})^2 = 2,\ \sum(x-\bar{x})(y-\bar{y}) = 0$

$$r = \frac{0}{\sqrt{2(2)}} = 0$$

b   If y = 1, when x = 6, then r = .509.
(Comment: Any y value greater than .973 will work.)

c   If y = −1, when x = 6, then r = −.509.
(Comment: any y value less than −.973 will work).

5.67   **a**

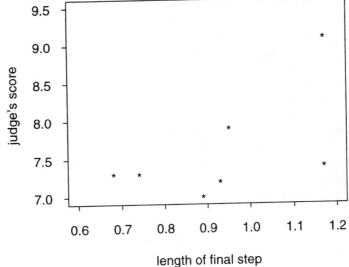

b $\sum x^2 - \dfrac{(\sum x)^2}{n} = .2157$, $\sum y^2 - \dfrac{(\sum y)^2}{n} = 3.08$, $\sum xy - \dfrac{(\sum x)(\sum y)}{n} = 0.474$

$b = \dfrac{.474}{.2157} = 2.1975$

$a = 7.6 - (2.1975)(.93286) = 5.550$

The least squares line is $\hat{y} = 5.550 + 2.1975 x$

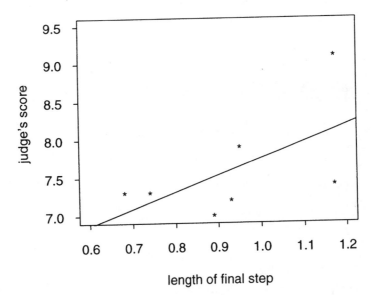

c $s_x = \sqrt{.2157/6} = .1896, \quad s_y = \sqrt{3.08/6} = .7165$

$$r = \frac{.474}{6(.1896)(.7165)} = .5815$$

This value of r suggests a moderate to weak linear relationship between x and y.

d

x rank	y rank	(x rank)(y rank)
6.5	5.0	32.5
6.5	7.0	45.5
4.0	2.0	8.0
3.0	1.0	3.0
1.0	3.5	3.5
2.0	3.5	7.0
5.0	6.0	30.0
		129.5

$$r_s = \frac{129.5 - \dfrac{7(8)^2}{4}}{\dfrac{7(6)(8)}{12}} = \frac{17.5}{28} = .625$$

This value is very close to the value of r in part **c**.

5.69   **a** $\sum x^2 - \dfrac{(\sum x)^2}{n} = 10, \ \sum y^2 - \dfrac{(\sum y)^2}{n} = 374, \ \sum xy - \dfrac{(\sum x)(\sum y)}{n} = -60$

$s_x = \sqrt{10/4} = 1.5811, \quad s_y = \sqrt{374/4} = 9.6695$

$$r = \frac{-60}{4(1.5811)(9.6695)} = -.981$$

This value of r suggests a strong negative linear relationship.

b

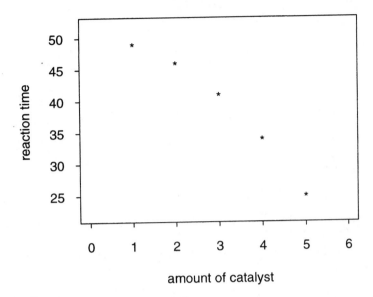

amount of catalyst

This plot suggests a curved relationship rather than a linear relationship.

5.71　　**a**　　$\text{SSResid} = \sum y^2 - a \sum y - b \sum xy = \sum y^2 - (\bar{y} - b\bar{x}) \sum y - b \sum xy$

$= \sum y^2 - \bar{y} \sum y + b\bar{x} \sum y - b \sum xy = (\sum y^2 - n\bar{y}^2) - b(\sum xy - n\overline{xy})$

$= \text{SSTo} - b(\text{numerator of b})$

　　　　　b　　Since b and the (numerator of b) have the same sign, their product must be a non-negative number. Therefore, SSResid = SSTo − (a non-negative number). Hence SSResid cannot be any larger than SSTo.

Chapter 6
Probability

Section 6.1

6.1   **a**   In the long run, 1% of all people who suffer cardiac arrest in New York City survive.

        b   1% of 2329 is .01(2329) = 23.29. So in this study there must have been only 23 or 24 survivors.

6.3   The outcomes *selected student is a recent immigrant* and *selected student has TB* are dependent events. Knowing that the selected student is a recent immigrant increases the probability of having TB from .0006 to .0075.

6.5   The outcome '*selected smoker who is trying to quit uses a nicotine aid*' and '*selected smoker who had attempted to quit begins smoking again within two weeks*' are dependent events. Knowing that the selected smoker used a nicotine aid reduces the probability of beginning smoking again within 2 weeks from 0.62 to 0.60.

6.7   **a**   Assuming the events are independent,

                P(Jeanie forgets all three errands)

                = P(Jeanie forgets first) x P(Jeanie forgets second) x P(Jeanie forgets third)

                = (0.1)(0.1)(0.1) = 0.001

        b   P(Jeanie remembers at least one)

                = 1 − P(Jeanie forgets all three) = 1 − .001 = .999

        c   P(Jeanie remembers the first errand but not the second or third)

                = P(Jeanie remembers first) x P(Jeanie forgets second) x P(Jeanie forgets third)

                = (1 − 0.1)(0.1)(0.1) = (0.9)(0.1)(0.1) = 0.009

6.9   **a** $P(\text{Visa}) = \dfrac{7000}{10000} = .7$

b   P(student has both cards) = $\dfrac{5000}{10000} = .5$

c $\dfrac{5000}{7000} = \dfrac{5}{7} = .714286$

d   No, because knowing that the student has a Visa card changes (increases) the probability that the student has a Mastercard from .6 to .714286.

e   Knowing the student has a Visa card, the probability that this student has both type of cards is $\dfrac{4200}{7000} = .6$. The events *has a Visa card* and *has a Mastercard* are independent since the probability of having a Mastercard does not change from .6.

6.11 **a**   The expert assumed that the positions of the two valves were independent.

b   If the car was driven in a straight line, the relative positions of the valves would remain unchanged. Thus, the probability of them ending up at a one o'clock, six o'clock position would be 1/12, not 1/144. The value 1/144 is smaller than the correct probability of occurrence.

6.13 **a**   P(1–2 subsystem works) = P(1 works) · P(2 works) = (.9)(.9) = .81

b   P(1–2 subsystem doesn't work) = 1 – P(1–2 subsystem works) = 1 – .81 = .19

P(3–4 subsystem doesn't work) = P(1–2 subsystem doesn't work) = .19

c   P(system won't work)

= P(1–2 subsystem doesn't work) x P(3–4 subsystem doesn't work)

= (.19)(.19) = .0361

P(system will work) = 1 – .0361 = .9639

d   P(system won't work) = (.19)(.19)(.19) = .006859

P(system will work) = 1 – .006859 = .993141

e   P(1–2 subsystem works) = (.9)(.9)(.9) = .729

P(1–2 subsystem won't work) = 1 – .729 = .271

P(system won't work) = (.271)(.271) = .073441

P(system works) = 1 – .073441 = .926559

Section 6.2

6.15 **a** .10 + .09 + .06 + .05 + .07 + .05 + .03 + .01 = .46

b It is more likely that they were in the first priority group than in the fourth priority group. (.10 compared to .05)

c If you are in the third priority group next term, the probability that you get more than 9 units during the first call is .06 + .03 = .09.

Section 6.3

6.17 **a**

		High School GPA		
		2.5 –< 3.0	3.0 –< 3.5	3.5 and above
Probation	Yes	.10	.11	.06
	No	.09	.27	.37

b .10 + .11 + .06 = .27

c .06 + .37 = .43

d P(selected student has a GPA of 3.5 or above and is on probation) = .06

P(selected student has a GPA of 3.5 or above) = .43

P(selected student is on probation) = .27

Since .43(.27) = .1161 does not equal .06, the events under question are not independent.

e 50/95 = .5263

f 30/215 = .1395

6.19 Total Monterey County = 369,000
Total San Luis Obispo County = 236,000
Total Santa Barbara County = 392,000
Total Ventura County = 736,000

Total Count = 1,733,000

Total Caucasian = 1,003,000
Total Hispanic = 528,000
Total Black = 61,000
Total Asian = 122,000
Total American Indian = 19,000

a $\dfrac{736,000}{1,733,000} = 0.4247$

b $\dfrac{231,000}{736,000} = 0.3139$

c $\qquad \dfrac{231,000}{528,000} = 0.4375$

d $\qquad \dfrac{9000}{1,733,000} = 0.0052$

e $\qquad$ P (Asian or from San Luis Obispo) = $\dfrac{122000}{1,733,000} + \dfrac{236000}{1,733,000} - \dfrac{9000}{1,733,000} = 0.2014$

f $\qquad$ P (Asian or from San Luis Obispo, but not both) =
$$\dfrac{122000}{1,733,000} + \dfrac{236000}{1,733,000} - \dfrac{18000}{1,733,000} = 0.1962$$

g $\qquad$ P (both Caucasian) = $(\dfrac{1,003,000}{1,733,000})(\dfrac{1,002,999}{1,732,999}) = 0.335$

h $\qquad$ P (neither Caucasian) = $(\dfrac{730000}{1733000})(\dfrac{729999}{1732999}) = 0.1774$

i $\qquad$ P (exactly one Caucasian) = P (first is Caucasian, second is not) + P (first is not Caucasian, second is Caucasian) =
$$(\dfrac{1003,000}{1733000})(\dfrac{730000}{1732999}) + (\dfrac{730000}{1733000})(\dfrac{1003000}{1732999}) = .2438 + .2438 = 0.4876$$

j $\qquad$ P (both from same county) = P (both from Monterey) + P (both from San Luis Obispo) + P (both from Santa Barbara) + P (both from Ventura) =
$$(\dfrac{369,000}{1,733,000})(\dfrac{368,999}{1,732,999}) + (\dfrac{236,000}{1,733,000})(\dfrac{235,999}{1732999}) + (\dfrac{392000}{1733000})(\dfrac{391999}{1732999}) + (\dfrac{736000}{1733000})(\dfrac{735999}{1732999})$$
$$= 0.0453 + 0.0185 + 0.0512 + 0.1804 = 0.2954$$

k $\qquad$ P (both from different racial/ethnic group) = 1 – P (both from the same racial/ethnic group = 1 – P (both Caucasian) – P (both Hispanic) – P (both Black) – P (both Asian) – P (both American Indian)

$\qquad$ P (both Caucasian) = 0.335, P (both Hispanic) = $(\dfrac{528,000}{1733000})(\dfrac{527999}{1732999}) = 0.0928$

$\qquad$ P (both Black) = $(\dfrac{61000}{1733000})(\dfrac{60999}{1732999}) = 0.0012$

$\qquad$ P (both Asian) = $(\dfrac{122000}{1733000})(\dfrac{121999}{1732999}) = 0.0050$

$\qquad$ P (both American Indian) = $(\dfrac{19000}{1733000})(\dfrac{18999}{1732999}) = 0.0001$

$\qquad$ P (both from different racial/ethnic groups) =
$\qquad$ 1- 0.335 – 0.0928 – 0.0012 – 0.0050 – 0.0001 = 0.5659

6.21   **a**   Assume that the requests are numbered from 1 to 20 as follows:

> 1–5   requests for a single license
> 6–14   requests for 2 licenses
> 15–20   requests for 3 licenses

Select a two digit random number from the table. If it is over 20 ignore it. If it is a number from 1 to 20 retain it. If it is a number from 1 to 5 then 1 license will be granted on this selection. If it is a number from 6–14 then two licenses will be granted for this selected number. If it is a number from 15–20 then three licenses will be granted on this selected number.

Select a second random number between 1 and 20 and grant the license request.

Select a third random number between 1 and 20 and grant the license request.

Select a fourth random number between 1 and 20 and grant the request if there are enough licenses remaining to satisfy the request.

Repeat the previous step until all 10 licenses have been granted.

The simulation results will vary from one simulation to another. The approximate probability obtained from a simulation with 20,000 trials done by computer was .3467.

b   An alternative procedure for distributing the licenses might be as follows. The twenty individuals requested a total of 41 licenses. Create a box containing 41 slips of paper. If an individual requested 1 license, his name is placed on only 1 piece of paper. If an individual requested two licenses, his name is placed on two slips. If he requested three licenses, his name is placed on three slips. Choose 10 slips of paper at random from the collection. Licenses are granted to those individuals whose names are selected. A person is granted as many licenses as the number of times their name is selected.

6.23   The simulation results will vary from one simulation to another. The approximate probability should be around .8468.

Supplementary Exercises

6.25   **a**   P(Dreyer's ice cream is purchased) = .20 + .25 = .45

b   P(Von's brand is not purchased) = .10 + .15 + .20 + .25 = .70

c   P(size purchased is larger than a pint) = .20 + .25 + .30 = .75

6.27   **a** $P(\text{twins}) = \dfrac{500,000}{42,005,100} = .0119$

b $P(\text{quadruplets}) = \dfrac{100}{42,005,100} = .00000238$

c $P(\text{more than a single child}) = \dfrac{505,100}{42,005,100} = .012$

6.29 The simulation results will vary from one simulation to another. The approximate probability obtained from a simulation with 20,000 trials done by computer was .8975.

Chapter 7
Population Distributions

Section 7.1

7.1   **a**   discrete

     b   continuous

     c   discrete

     d   discrete

     e   continuous

7.3   **a**

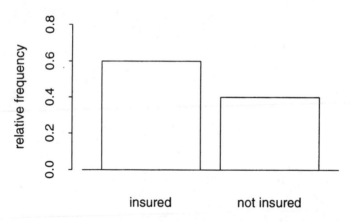

     b   P(individual does not have earthquake insurance) = 0.4

7.5   **a**

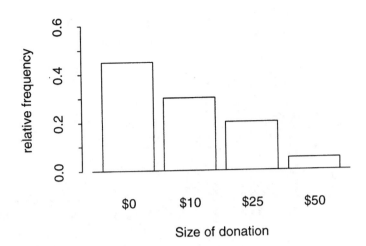

b   x = 0, since its probability is the largest.

c $P(x \geq 25) = P(x = 25) + P(x = 50) = 0.20 + 0.05 = 0.25$

d $P(x > 0) = 1 - P(x = 0) = 1 - 0.45 = 0.55$

7.7   **a** $P(x \leq 100) = 0.05 + 0.10 + 0.12 + 0.14 + 0.24 + 0.17 = 0.82$

b $P(x > 100) = 1 - P(x \leq 100) = 1 - 0.82 = 0.18$

c $P(x \leq 99) = 0.05 + 0.10 + 0.12 + 0.14 + 0.24 = 0.65$
$P(x \leq 97) = 0.05 + 0.10 + 0.12 = 0.27$

7.9   **a**   Supplier 1

b   Supplier 2

c   Supplier 1, because the bulbs generally last longer and have less variability in their lifetimes.

d   About 1000

e   About 100

Section 7.2

7.11   **a**

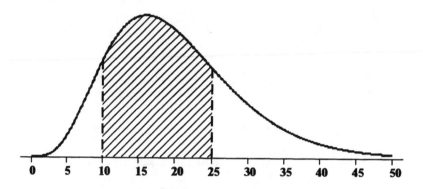

b

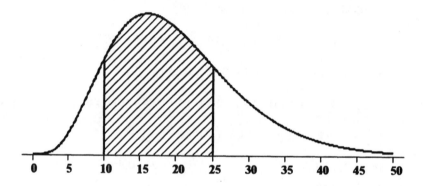

c

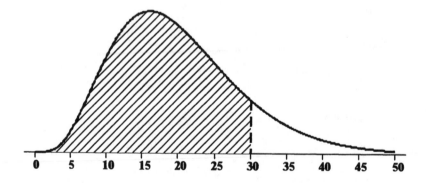

d

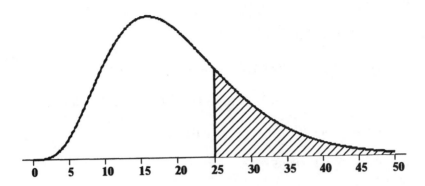

e

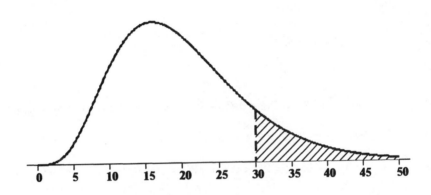

7.13 **a**   Total area = 1/2(0.40)(5) = 1

b   P(x < 0.20) = 1/2(0.20)(5) = 0.5
P(x < 0.10) = 1/2(0.10)(5/2) = 0.125
P(x > 0.30) = 1/2(0.10)(5/2) = 0.125

c   P(0.10 ≤ x ≤ 0.20) = 1 − P(x < 0.10 or x > 0.20)
From part **b**, P(x < 0.10) = 0.125 and P(x > 0.20) = 0.5
Hence, P(0.10 ≤ x ≤ 0.20) = 1 − (0.125 + 0.5) = 0.325

d   P(0.15 ≤ x ≤ 0.25) = 1 − [P(x ≤ 0.15) + P(x ≥ 0.25)]

$$= 1 - \left[\frac{1}{2}(.15)(5)(.75) + \frac{1}{2}(.4-.25)(5)(.75)\right]$$

= 1 − [0.28125 + 0.28125] = 1 − 0.5625 = 0.4375

7.15 **a**   The area under the density curve must equal 1. Since this area is a triangle, we have
½(1)h = 1. So h = 2.

b   P(x > 0.5) = (1/2)(0.5)(1) = 0.25
c   P(x ≤ 0.25) = 1 − P(x > 0.25) = 1 − [(1/2)(1 − 0.25)(1.5)] = 1 − 9/16 = 7/16

Section 7.3

7.17 **a** $P(z < -1.28) = 0.1003$

b $P(z > 1.28) = 1 - P(z \leq 1.28) = 1 - 0.8997 = 0.1003$

c $P(-1 < z < 2) = P(z < 2) - P(z < -1) = 0.9772 - 0.1587 = 0.8185$

d $P(z > 0) = 1 - P(z \leq 0) = 1 - 0.5 = 0.5$

e $P(z > -5) = 1 - P(z \leq -5) \approx 1 - 0 = 1$

f $P(-1.6 < z < 2.5) = P(z < 2.5) - P(z < -1.6) = 0.9938 - 0.0548 = 0.9390$

g $P(z < 0.23) = 0.5910$

7.19 **a** $P(z < 0.1) = 0.5398$

b $P(z < -0.1) = 0.4602$

c $P(0.40 < z < 0.85) = P(z < 0.85) - P(z < 0.4) = 0.8023 - 0.6554 = 0.1469$

d $P(-0.85 < z < -0.40) = P(z < -0.4) - P(z < -0.85) = 0.3446 - 0.1977 = 0.1469$

e $P(-0.40 < z < 0.85) = P(z < 0.85) - P(z < -0.4) = 0.8023 - 0.3446 = 0.4577$

f $P(-1.25 < z) = 1 - P(z \leq -1.25) = 1 - 0.1056 = 0.8944$

g $P(z < -1.5 \text{ or } z > 2.5) = P(z < -1.5) + 1 - P(z \leq 2.5) = 0.0668 + 1 - 0.9938 = 0.0730$

7.21 **a** $P(z > z^*) = 0.03 \implies z^* = 1.88$

b $P(z > z^*) = 0.01 \implies z^* = 2.33$

c $P(z < z^*) = 0.04 \implies z^* = -1.75$

d $P(z < z^*) = 0.10 \implies z^* = -1.28$

7.23 **a** 91st percentile = 1.34

b 77th percentile = 0.74

c 50th percentile = 0

d 9th percentile = -1.34

e They are negatives of one another. The 100pth and 100(1-p)th percentiles will be negatives of one another, because the z curve is symmetric about 0.

7.25 **a** $P(x > 4000) = P(z > \dfrac{4000 - 3432}{482}) = P(z > 1.1784) = 0.1193$

$P(3000 \le x \le 4000) = P(\dfrac{3000 - 3432}{482} \le z \le \dfrac{4000 - 3432}{482})$

$= P(-0.8963 \le z \le 1.1784) = P(z \le 1.1784) - P(z < -0.8963) = 0.8807 - 0.1851$
$= 0.6956$

b $P(x < 2000) + P(x > 5000) = P(z < \dfrac{2000 - 3432}{482}) + P(z > \dfrac{5000 - 3432}{482})$

$= P(z < -2.97095) + P(z > 3.25311) = 0.0015 + 0.00057 = 0.00207$

c $P(x > 7 \text{ lbs}) = P(x > 7(453.6) \text{ grams}) = P(x > 3175.2) = P(z > \dfrac{3175.2 - 3432}{482})$

$= P(z > -0.53278) = 0.70291$

d   We find x_1^* and x_2^* such that $P(x < x_1^*) = 0.0005$ and $P(x > x_2^*) = 0.0005$. The most extreme 0.1% of all birthweights would then be characterized as weights less than x_1^* or weights greater than x_2^*. $P(x < x_1^*) = P(z < z_1^*) = 0.0005$ implies that $z_1^* = -3.2905$. So $x_1^* = \mu + z_1^* \sigma = 3432 + 482(-3.2905) = 1846$ grams. $P(x > x_1^*) = P(z > z_2^*) = 0.0005$ implies that $z_2^* = 3.2905$. So $x_2^* = \mu + z_2^* \sigma = 3432 + 482(3.2905) = 5018$ grams. Hence the most extreme 0.1% of all birthweights correspond to weights less than 1846 grams or weights greater than 5018 grams.

e   If x is a random variable with a normal distribution and a is a numerical constant (not equal to 0) then $y = ax$ also has a normal distribution. Furthermore,

   mean of $y = a \times$ (mean of x)

and

   standard deviation of $y = a \times$ (standard deviation of x).

Let y be the birthweights measured in pounds. Recalling that one pound = 453.6

grams, we have $y = \left(\dfrac{1}{453.6}\right) x$, so $a = \dfrac{1}{453.6}$. The distribution of y is normal with

mean equal to $3432/453.6 = 7.56614$ pounds and standard deviation equal to $482/453.6$

$= 1.06261$ pounds. So $P(y > 7 \text{ lbs}) = P(z > \dfrac{7 - 7.56614}{1.06261}) = P(z > -0.53278) = 0.70291$.

As expected, this is the same answer that we obtained in part **c**.

7.27   For the second machine,

P(that a cork doesn't meet specifications) $= 1 - P(2.9 \le x \le 3.1)$

$= 1 - P(\dfrac{2.9 - 3.05}{0.01} \le z \le \dfrac{3.1 - 3.05}{0.01}) = 1 - P(-15 \le z \le 5) \approx 1 - 1 = 0.$

7.29 $\dfrac{c - 120}{120} = -1.28 \implies c = 120 - 1.28(20) = 94.4$

Task times of 94.4 seconds or less qualify an individual for the training.

7.31   **a** $P(x \le 60) = P(z \le \dfrac{60-60}{15}) = P(z \le 0) = 0.5$

$P(x < 60) = P(z < 0) = 0.5$

b $P(45 < x < 90) = P(\dfrac{45-60}{15} < z < \dfrac{90-60}{15})$

$= P(-1 < z < 2) = 0.9772 - 0.1587 = 0.8185$

c $P(x \ge 105) = P(z \ge \dfrac{105-60}{15}) = P(z \ge 3) = 1 - P(z < 3) = 1 - 0.9987 = 0.0013$

The probability of a typist in this population having a net rate in excess of 105 is only 0.0013. Hence it would be surprising if a randomly selected typist from this population had a net rate in excess of 105.

d $P(x > 75) = P(z > \dfrac{75-60}{15}) = P(z > 1) = 1 - P(z \le 1) = 1 - 0.8413 = 0.1587$

$P(\text{both exceed } 75) = (0.1587)(0.1587) = 0.0252$

e $P(z < z^*) = 0.20 \Rightarrow z^* = -0.84$

$x^* = \mu + z^*\sigma \Rightarrow x^* = 60 + (-0.84)(15) = 60 - 12.6 = 47.4$

So typing speeds of 47.4 words or less per minute would qualify individuals for this training.

Section 7.4

7.33   Since this plot appears to be very much like a straight line, it is reasonable to conclude that the normal distribution provides an adequate description of the steam rate distribution.

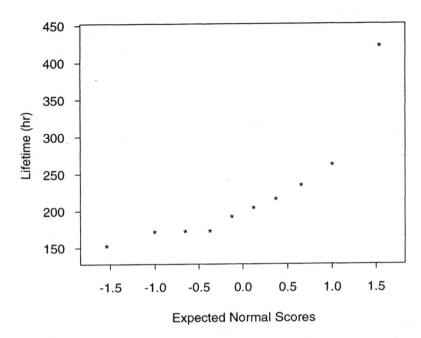

Since the graph exhibits a pattern substantially different from that of a straight line, one would conclude that the distribution of the variable "component lifetime" cannot be adequately modeled by a normal distribution. It is worthwhile noting that this "deviation from normality" could be due to the single outlying value of 422.6.

7.37

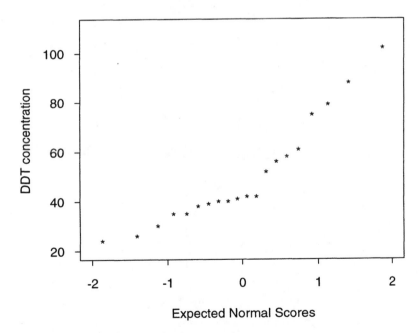

Although the graph follows a straight line pattern approximately, there is a distinct "kink" in the graph at about the value 45 on the vertical axis. Points corresponding to DDT concentration less than 45 seem to follow one straight line pattern while those to the right of 45 seem to follow a different straight line pattern. A normal distribution may not be an appropriate model for this population.

7.39 Histograms of the square root transformed data as well as the cube root transformed data are given below. It appears that the histogram of the cube root transformed data is more symmetric than the histogram of the square root transformed data. (However, keep in mind that the shapes of these histograms are somewhat dependent on the choice of class intervals.)

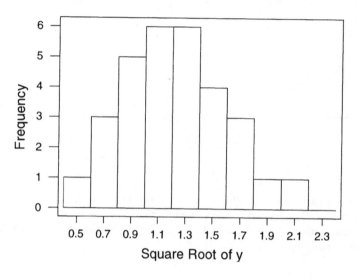

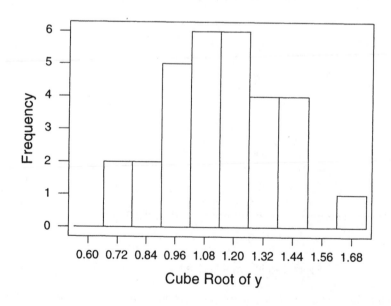

7.41 **a**   The required frequency distribution is given below.

class	frequency	relative frequency
0-<100	22	0.22
100-<200	32	0.32
200-<300	26	0.26
300-<400	11	0.11
400-<500	4	0.04
500-<600	3	0.03
600-<700	1	0.01
700-<800	0	0
800-<900	1	0.01

b

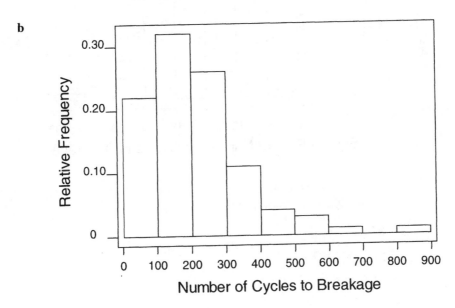

This histogram is positively skewed.

c   Data were square root transformed and the corresponding relative frequency histogram is shown below. Clearly the distribution of the transformed data is more symmetric than that of the original data.

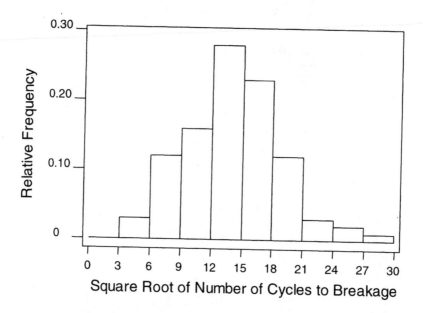

7.43   **a**

0	56667889999
1	000011122456679999
2	0011111368
3	11246
4	18
.5	
6	8
7	
8	
9	39 HI: 448

b   A density histogram of the body mass data is shown below. It is positively skewed. The scale on the horizontal axis extends upto body mass = 500 in order to accommodate the HI value of 448.

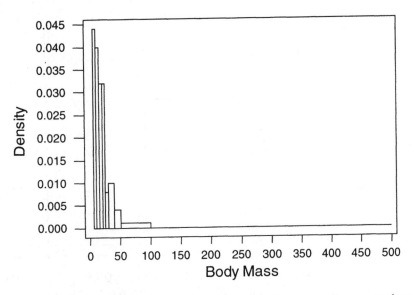

If statistical inference procedures based on normality assumptions are to be used to draw conclusions from the body mass data of this problem, then a transformation should be considered so that, on the transformed scale, the data are approximately normally distributed.

c

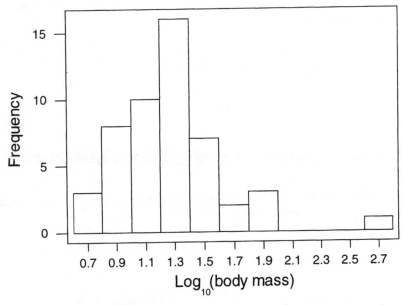

The normal curve fits the histogram of the log transformed data better than the histogram of the original data.

d

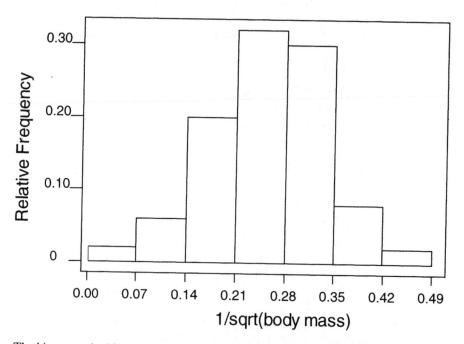

The histogram in this transformed scale does appear to have an approximate bell-shape, i.e., in the transformed scale, the data do appear to follow a normal distribution model.

Supplementary Exercises

7.45 $P(x < 4.9) = P(z < \dfrac{4.9 - 5}{0.05}) = P(z < -2) = 0.0228$

   $P(5.2 < x) = P(\dfrac{5.2 - 5}{0.05} < z) = P(4 < z) \approx 0$

7.47   **a** $P(x < 5'7'') = P(x < 67'') = P(z < \dfrac{67 - 66}{2}) = P(z < 0.5) = 0.6915$

No, the claim that 94% of all women are shorter than 5'7" is not correct. Only about 69% of all women are shorter than 5'7".

   b   About 69% of adult women would be excluded from employment due to the height requirement.

7.49   **a** $P(5.9 < x < 6.15) = P(\dfrac{5.9 - 6}{0.1} < z < \dfrac{6.15 - 6}{0.1}) = P(-1 < z < 1.5)$
   $= 0.9332 - 0.1587 = 0.7745$

b $P(6.1 < x) = P(\dfrac{6.1-6}{0.1} < z) = P(1 < z) = 1 - P(z \le 1) = 1 - 0.8413 = 0.1587$

c $P(x < 5.95) = P(z < \dfrac{5.95-6}{0.1}) = P(z < -0.5) = 0.3085$

d   The largest 5% of the pH values are those pH values which exceed the 95th percentile.

The 95th percentile is $6 + 1.645(0.1) = 6.1645$.

7.51 **a** $P(250 < x < 300) = P((250 - 266)/16 < z < (300 - 266)/16)$
$= P(-1 < z < 2.13) = 0.9834 - 0.1587 = 0.8247$

b $P(x < 240) = P(z < (240 - 266)/16) = P(z < -1.63) = 0.0516$

c $P(x \text{ is within 16 days of the mean duration})$
$= P(250 < x < 282) = P(-1 < z < 1) = 0.8413 - 0.1587 = 0.6826$

d $P(310 \le x) = P((310 - 266)/16 < z) = P(2.75 < z)$
$= 1 - P(z \le 2.75) = 1 - 0.9970 = 0.0030$

The chances of a pregnancy having a duration of at least 310 days is 0.003. This is a small value, so there is a bit of skepticism concerning the claim of this lady.

e   If the duration is 261 days or less, then the date of birth will be $261 + 14 = 275$ days or less after coverage began. Hence the insurance company will not pay the benefits.

$P(x < 261) = P(z < (261 - 266)/16) = P(z < -0.31) = 0.3783 \approx 38\%$

When date of conception is 14 days after coverage began, about 38% of the time the insurance company will refuse to pay the benefits because of the 275 day requirement.

Chapter 8
Sampling Variability and Sampling Distributions

Section 8.1

8.1   A statistic is any quantity computed from the observations in a sample. A population characteristic is a quantity which describes the population from which the sample was taken.

8.3   **a**   population characteristic
      b   statistic
      c   population characteristic
      d   statistic
      e   statistic

8.5

We selected a random sample of size n=2 from the population of Exercise 8.4 fifty times and calculated the sample mean for each sample. The results are shown in the table below. Your own simulation may produce different results.

Trial number	First sample value	Second sample value	Sample mean	Trial number	First sample value	Second sample value	Sample mean
1	4	1	2.5	26	3	2	2.5
2	4	3	3.5	27	3	1	2.0
3	2	4	3.0	28	2	3	2.5
4	4	2	3.0	29	2	1	1.5
5	3	1	2.0	30	4	2	3.0
6	4	3	3.5	31	4	1	2.5
7	3	1	2.0	32	2	3	2.5
8	4	2	3.0	33	4	3	3.5
9	2	3	2.5	34	3	2	2.5
10	1	4	2.5	35	4	1	2.5
11	4	3	3.5	36	1	4	2.5
12	1	4	2.5	37	1	4	2.5
13	3	1	2.0	38	1	2	1.5
14	3	4	3.5	39	4	1	2.5
15	2	3	2.5	40	2	4	3.0
16	3	2	2.5	41	4	3	3.5
17	4	2	3.0	42	3	2	2.5
18	2	3	2.5	43	2	4	3.0
19	3	2	2.5	44	1	2	1.5
20	1	2	1.5	45	1	4	2.5
21	4	3	3.5	46	3	4	3.5
22	2	1	1.5	47	3	4	3.5
23	3	2	2.5	48	1	2	1.5
24	4	1	2.5	49	4	1	2.5
25	2	1	1.5	50	4	1	2.5

The relative frequency distribution of the 50 simulated values of $\overline{x}$ and the corresponding probabilities from the sampling distribution of $\overline{x}$ are as follows.

Value of $\overline{x}$	Relative frequency from simulation	Probability from sampling distribution
1.5	0.14	0.167
2	0.08	0.167
2.5	0.46	0.333
3	0.14	0.167
3.5	0.18	0.167

The relative frequencies from the simulation are approximately equal to the corresponding probabilities from the sampling distribution of $\overline{x}$. We expect the approximations to improve as the number of simulations increases. Your own simulations may lead to somewhat different results.

8.7 **a**

Sample	Value of t	Sample	Value of t
1, 5	6	5, 10	15
1, 10	11	5, 20	25
1, 20	21	10, 20	30

Value of t	6	11	15	21	25	30
Probability	1/6	1/6	1/6	1/6	1/6	1/6

b The population mean is $\mu = (1 + 5 + 10 + 20)/4 = 36/4 = 9$

$\mu_t = (1/6)(6) + (1/6)(11) + (1/6)(15) + (1/6)(21) + (1/6)(25) + (1/6)(30) = 108/6 = 18$

μ_t is twice the value of μ. More generally, the value of μ_t will equal the value of μ times the sample size.

8.9 **Note:** Statistic #3 is sometimes called the **midrange**.

Sample	Value of Mean	Value of Median	Value of statistic #3 (midrange)
2, 3, 3*	2.67	3	2.5
2, 3, 4	3.00	3	3.0
2, 3, 4*	3.00	3	3.0
2, 3*, 4	3.00	3	3.0
2, 3*, 4*	3.00	3	3.0
2, 4, 4*	3.33	4	3.0
3, 3*, 4	3.33	3	3.5
3, 3*, 4*	3.33	3	3.5
3, 4, 4*	3.67	4	3.5
3*, 4, 4*	3.67	4	3.5

Sampling distribution of statistic #1.

Value of $\overline{x}$	2.67	3	3.33	3.67
Probability	0.1	0.4	0.3	0.2

Sampling distribution of statistic #2.

Value of median	3	4
Probability	0.7	0.3

Sampling distribution of statistic #3.

Value of midrange	2.5	3	3.5
Probability	0.1	0.5	0.4

Some of the many points to be considered when selecting an estimator of a population parameter are listed below.

(i)   Is the estimator unbiased? i.e., is the mean of the sampling distribution of the estimator equal to the parameter being estimated?

(ii)   Is there a high probability that the value of the estimator will be "sufficiently close" to the true parameter value?

(iii)   Is the estimator resistant to the influence of outliers?

Your own choice of a statistic may be different from those of others.

8.11   **a** $\mu = \dfrac{8 + 14 + 16 + 10 + 11}{5} = \dfrac{59}{5} = 11.8$

    b   One possible random sample consists of elements 1 and 4, whose values are 8 and 10. The value of $\bar{x}$ for this sample is $(8 + 10)/2 = 9$.

    c   Twenty-four additional samples of size 2 and their means are:

Sample	Sample mean	Sample	Sample mean
14, 16	15.0	10, 16	13.0
16, 14	15.0	10, 11	10.5
11, 14	12.5	11, 14	12.5
11, 10	10.5	16, 8	12.0
14, 10	12.0	16, 8	12.0
16, 11	13.5	8, 10	9.0
14, 10	12.0	11, 10	10.5
8, 11	9.5	8, 10	9.0
16, 14	15.0	10, 8	9.0
16, 10	13.0	8, 10	9.0
10, 11	10.5	16, 14	15.0
8, 14	11.0	10, 16	13.0

d

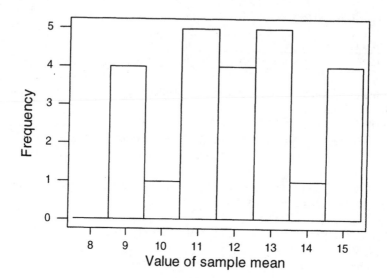

For this group of 25 samples of size two, the sample means are either close to the population mean, or at either extreme. The sample means tend to differ quite a bit from one sample to the next.

8.13   If samples of size ten rather than size five had been used, the histograms would be similar in that they both would be centered close to 260.25. They would differ in that the histogram based on n = 10 would have less variability than the histogram based on n = 5 (see Figure 8.4 of the text).

Section 8.2

8.15   For n = 36, 50, 100, and 400

8.17   **a** $\mu_{\bar{x}} = 40, \sigma_{\bar{x}} = \dfrac{5}{\sqrt{64}} = \dfrac{5}{8} = 0.625$

Since n = 64, which exceeds 30, the shape of the sampling distribution will be approximately normal.

b $P(\mu - 0.5 < \bar{x} < \mu + 0.5) = P(39.5 < \bar{x} < 40.5) = P(\dfrac{39.5 - 40}{0.625} < z < \dfrac{40.5 - 40}{0.625})$

$= P(-0.8 < z < 0.8) = 0.7881 - 0.2119 = 0.5762$

c $P(\bar{x} < 39.3 \text{ or } \bar{x} > 40.7) = P(z < \dfrac{39.3 - 40}{0.625} \text{ or } z > \dfrac{40.7 - 40}{0.625})$

$= 1 - P(-1.12 < z < 1.12) = 1 - [0.8686 - 0.1314] = 0.2628$

8.19   **a** $\mu_{\bar{x}} = 2 \text{ and } \sigma_{\bar{x}} = \dfrac{0.8}{\sqrt{9}} = 0.267$

b   For n = 20, $\mu_{\bar{x}} = 2$ and $\sigma_{\bar{x}} = \dfrac{0.8}{\sqrt{20}} = 0.179$

For n = 100, $\mu_{\bar{x}} = 2$ and $\sigma_{\bar{x}} = \dfrac{0.8}{\sqrt{100}} = 0.08$

In all three cases $\mu_{\bar{x}}$ has the same value, but the standard deviation of $\bar{x}$ decreases as n increases. A sample of size 100 would be most likely to result in an $\bar{x}$ value close to μ. This is because the sampling distribution of $\bar{x}$ for n = 100 has less variability than those for n = 9 or 20.

8.21   **a** $\sigma_{\bar{x}} = \dfrac{5}{\sqrt{25}} = 1$

$P(64 \le \bar{x} \le 67) = P(\dfrac{64-65}{1} \le z \le \dfrac{67-65}{1}) = P(-1 \le z \le 2)$

$= 0.9772 - 0.1587 = 0.8185$

$P(68 \le \bar{x}) = P(\dfrac{68-65}{1} \le z) = P(3 \le z) = 1 - P(z < 3) = 1 - 0.9987 = 0.0013$

b   Because the sample size is large we can use the normal approximation for the sampling distribution of $\bar{x}$ by the central limit theorem. We have

$\sigma_{\bar{x}} = \dfrac{5}{\sqrt{100}} = .5$

$P(64 \le \bar{x} \le 67) = P((64-65)/0.5 < z < (67-65)/0.5) = P(-2 < z < 4)$
$= 1 - 0.0228 = 0.9772$

$P(68 \le \bar{x}) = P((68-65)/.5 \le z) = P(6 \le z) = 1 - P(z < 6) \approx 1 - 1 = 0$

8.23   If the true process mean is equal to 0.5 in, then $\mu_{\bar{x}} = 0.5, \sigma_{\bar{x}} = \dfrac{0.02}{\sqrt{36}} = 0.00333$

P(the line will be shut down unnecessarily) $= 1 - P(0.49 \le \bar{x} \le 0.51)$

$= 1 - P(\dfrac{0.49-0.50}{0.00333} \le z \le \dfrac{0.51-0.50}{0.00333}) = 1 - P(-3 \le z \le 3) = 1 - (0.9987 - 0.0013) = 0.0026$

8.25   The total weight of their baggage will exceed the limit, if the average weight exceeds 6000/100 = 60.

$\mu_{\bar{x}} = 50, \ \sigma_{\bar{x}} = \dfrac{20}{\sqrt{100}} = 2 ,$

$P(6000 \le \text{total weight}) = P(60 \le \bar{x}) = P(\dfrac{60-50}{2} \le z) = P(5 \le z) = 1 - P(z < 5) \approx 1 - 1 = 0.$

Section 8.3

8.27 **a** $\mu_p = 0.65, \sigma_p = \sqrt{0.65(0.35)/10} = 0.15083$

b $\mu_p = 0.65, \sigma_p = \sqrt{0.65(0.35)/20} = 0.10665$

c $\mu_p = 0.65, \sigma_p = \sqrt{0.65(0.35)/30} = 0.08708$

d $\mu_p = 0.65, \sigma_p = \sqrt{0.65(0.35)/50} = 0.06745$

e $\mu_p = 0.65, \sigma_p = \sqrt{0.65(0.35)/100} = 0.04770$

f $\mu_p = 0.65, \sigma_p = \sqrt{0.65(0.35)/200} = 0.03373$

8.29 **a** $\mu_p = 0.005, \sigma_p = \sqrt{\dfrac{(0.005)(0.995)}{100}} = 0.007$

b   Since $n\pi = 100(0.005) = 0.5$ is less than 5, the sampling distribution of p cannot be approximated well by a normal curve.

c   The requirement is that $n\pi \geq 5$, which means that n would have to be at least 1000.

8.31 **a**   For $\pi = 0.5$, $\mu_p = 0.5$ and $\sigma_p = \sqrt{\dfrac{0.5(0.5)}{225}} = 0.0333$

For $\pi = 0.6$, $\mu_p = 0.6$ and $\sigma_p = \sqrt{\dfrac{0.6(0.4)}{225}} = 0.0327$

For both cases, $n\pi \geq 5$ and $n(1 - \pi) \geq 5$. Hence, in each instance, p would have an approximately normal distribution.

b   For $\pi = 0.5$, $P(p \geq 0.6) = P(z \geq \dfrac{0.6-0.5}{0.0333}) = P(z \geq 3) = 1 - P(z < 3)$

$= 1 - 0.9987 = 0.0013.$

For $\pi = 0.6$, $P(p \geq 0.6) = P(z \geq \dfrac{0.6-0.6}{0.0327}) = P(z \geq 0) = 1 - P(z < 0)$

$= 1 - 0.5000 = 0.5000$

c   When $\pi = 0.5$, the $P(p \geq 0.6)$ would decrease.

When $\pi = 0.6$, the $P(p \geq 0.6)$ would remain the same.

8.33   **a** $\mu_p = \pi = 0.05, \sigma_p = \sqrt{\dfrac{(0.05)(0.95)}{200}} = 0.01541$

$P(0.02 < p) = P(\dfrac{0.02 - 0.05}{0.01541} < z) = P(-1.95 < z)$

$= 1 - P(z \le -1.95) \approx 1 - 0.0258 = 0.9742$

b $\mu_p = \pi = 0.10, \sigma_p = \sqrt{\dfrac{(0.1)(0.9)}{200}} = 0.02121$

$P(p \le 0.02) = P(z \le \dfrac{0.02 - 0.10}{0.02121}) = P(z \le -3.77) \approx 0$

Supplementary Exercises

8.35   **a**   The sampling distribution of $\bar{x}$ will be approximately normal, with mean equal to 50

(lb) and standard deviation equal to $\dfrac{1}{\sqrt{100}} = 0.1$ (lb).

b $P(49.75 < \bar{x} < 50.25) = P(\dfrac{49.75 - 50}{0.1} < z < \dfrac{50.25 - 50}{0.1}) = P(-2.5 < z < 2.5)$

$= .9938 - .0062 = .9876$

c $P(\bar{x} < 50) = P(z < \dfrac{50 - 50}{0.1}) = P(z < 0) = 0.5$

8.37   **a** $\mu_{\bar{x}} = 52$ (minutes) and $\sigma_{\bar{x}} = \dfrac{2}{\sqrt{36}} = 0.33$ (minutes)

b $P(50 < \bar{x}) = P(\dfrac{50 - 52}{0.33} < z) = P(-6 < z) = 1 - P(z \le -6) \approx 1 - 0 = 1$

$P(55 < \bar{x}) = P(\dfrac{55 - 52}{0.33} < z) = P(9 < z) = 1 - P(z \le 9) \approx 1 - 1 = 0$

8.39 $\mu_p = 0.4, \sigma_p = \sqrt{\dfrac{(0.4)(0.6)}{100}} = 0.049$

a $P(0.5 \le p) \approx P(\dfrac{0.5 - 0.4}{0.049} \le \dfrac{p - 0.4}{0.049}) = P(2.04 \le z) = 1 - P(z < 2.04)$

$= 1 - 0.9793 = 0.0207$

b   The probability of obtaining a sample of 100 in which at least 60 participated in such a plan is almost 0 if $\pi = 0.40$. Hence, one would doubt that $\pi = 0.40$ and strongly suspect that π has a value greater than 0.4.

Chapter 9
Estimation Using a Single Sample

Section 9.1

9.1 Statistic II would be preferred because it is unbiased and has smaller variance than the other two.

9.3 The reported number of readings above 4 pCi is 68. The point estimate of π, the proportion of all homes on the reservation whose readings exceed 4 pCi is $p = \dfrac{68}{270} = 0.252$.

9.5 $p = \dfrac{245}{935} = 0.262$

9.7 **a** The value of σ will be estimated by using the statistic s. For this sample,

$$\Sigma x^2 = 1757.54, \Sigma x = 143.6, n = 12$$

$$s^2 = \frac{\Sigma x^2 - \dfrac{(\Sigma x)^2}{n}}{n-1} = \frac{1757.54 - \dfrac{(143.6)^2}{12}}{12-1} = \frac{1757.54 - 1718.4133}{11}$$

$$= \frac{39.1267}{11} = 3.557 \text{ and } s = \sqrt{3.557} = 1.886$$

b The population median will be estimated by the sample median. Since n = 12 is even, the sample median equals the average of the middle two values (6th and 7th values), i.e., $\dfrac{(11.3+11.4)}{2} = 11.35$.

c In this instance, a trimmed mean will be used. First arrange the data in increasing order. Then, trimming one observation from each end will yield an 8.3% trimmed mean. The trimmed mean equals $117.3/10 = 11.73$.

d   The point estimate of μ would be $\bar{x} = 11.967$. From part **a**, $s = 1.886$. Therefore the estimate of the 90th percentile is $11.967 + 1.28(1.886) = 14.381$.

9.9   **a** $\bar{x}_J = 120.6$

b   An estimate of the total amount of gas used by all these houses in January would be $10000(120.6) = 1,206,000$ therms. More generally, an estimate of the population total is obtained by multiplying the sample mean by the size of the population.

c $p = 8/10 = 0.8$

d   Using the sample median, an estimate of the population median usage is $(118 + 122)/2 = 120$ therms.

Section 9.2

9.11   The values for parts **a-d** are found in the table for Standard Normal Probabilities (Appendix Table II).

a   1.96

b   1.645

c   2.58

d   1.28

e   1.44 (approximately)

9.13   **a**   As the confidence level increases, the width of the confidence interval for π increases.

b   As the sample size increases, the width of the confidence interval for π decreases.

c   As the value of p gets farther from 0.5, either larger or smaller, the width of the confidence interval for π decreases.

9.15   The sample proportion p is $21/39 = 0.538$. The 95% confidence interval would be

$$0.5385 \pm 1.96\sqrt{\frac{0.5385(1-0.5385)}{39}} \Rightarrow 0.5385 \pm 0.1565 \Rightarrow (0.382, 0.695).$$

9.17   Based on the information given, a 95% confidence interval for π, the proportion of the population who felt that their financial situation had improved during the last year, is calculated as follows. $0.43 \pm 1.96\sqrt{\frac{0.43(1-0.43)}{930}} \Rightarrow 0.43 \pm 0.0318 \Rightarrow (0.398, 0.462).$

Hence, with 95% confidence, the percentage of people who felt their financial situation had improved during the last year is between 39.8% and 46.2%. These end points of the confidence interval are 3.2% away on either side of the estimated value of 43%. In the report, the value 3.2% has been rounded down to 3%. Thus, the claim of a 3% "margin of error" in the report is statistically justified.

An alternative way to justify the statement in the article is as follows. The mean of the sampling distribution of p is π and the standard deviation is equal to $\sqrt{\dfrac{\pi(1-\pi)}{930}}$. The largest possible standard deviation occurs when π is equal to 0.5, in which case the standard deviation is equal to 0.0164. Hence the probability that p will be within 0.03 (within 3 percent) of π is greater than or equal to P((-0.03/0.0164) < z < (0.03/0.0164)) = P(-1.8297 < z < 1.8297) = 0.933 which is approximately equal to 0.95. Hence the probability is 93.3% or greater that the value of p is within 3 percent of the true value of π.

9.19 A 90% confidence interval is $0.65 \pm 1.645\sqrt{\dfrac{0.65(1-0.65)}{150}} \Rightarrow 0.65 \pm 0.064 \Rightarrow (0.589, 0.714)$

Thus, we can be 90% confident that between 58.9% and 71.4% of Utah residents favor fluoridation. This is consistent with the statement that a clear majority of Utah residents favor fluoridation.

9.21 **a** $0.721 \pm 2.58\sqrt{\dfrac{0.721(0.279)}{500}} \Rightarrow 0.721 \pm 0.052 \Rightarrow (0.669, 0.773)$

b $0.279 \pm 1.96\sqrt{\dfrac{0.279(0.721)}{500}} \Rightarrow 0.279 \pm 0.039 \Rightarrow (0.24, 0.318)$

Based on this interval, we conclude that between 24% and 31.8% of college freshman are not attending their first choice of college.

c It would have been narrower.

9.23 **a** $p = \dfrac{442}{1005} = 0.4398$

The 90% confidence interval computed from the data of this sample is

$0.4398 \pm (1.645)\sqrt{\dfrac{0.4398(0.5602)}{1005}} \Rightarrow 0.4398 \pm 0.0258 \Rightarrow (0.4140, 0.4656)$.

b Since the entire interval is below 0.5, it is not plausible that a majority of U.S. adults feel that it makes no difference which party is in control.

9.25 The required sample size is $n = (0.25)\left[\dfrac{1.96}{0.02}\right]^2 = (0.25)(9604) = 2401$.

The recommended sample size is large because (1) no "ball-park" value for p is known and (2) the desired accuracy is quite small.

9.27 $n = 0.25\left[\dfrac{1.96}{B}\right]^2 = 0.25\left[\dfrac{1.96}{0.05}\right]^2 = 384.16;$ take n = 385.

Section 9.3

9.29   **a**   90%
         b   95%
         c   95%
         d   99%
         e   1%
         f   0.5%
         g   5%

9.31   **a** $\bar{x}$ is the midpoint of the interval. So, $\bar{x} = \dfrac{114.4 + 115.6}{2} = 115.0$.

         b   As the confidence level increases the width of the interval increases. Hence (114.4, 115.6) is the 90% interval and (114.1, 115.9) is the 99% interval.

9.33   **a**   The 90% confidence interval would have been narrower, since its z critical value would have been smaller.

         b   The statement is incorrect. The 95% refers to the percentage of *all possible* samples that result in an interval that includes μ, not to the chance (probability) that a specific interval contains μ.

         c   Again this statement is incorrect. While we would expect *approximately* 95 of the 100 intervals constructed to contain μ, we cannot be *certain* that exactly 95 out of 100 of them will. The 95% refers to the percentage of *all possible* intervals that include μ.

9.35   **a** $0.5 \pm 1.96\dfrac{0.4}{\sqrt{77}} \Rightarrow 0.5 \pm 0.089 \Rightarrow (0.411, 0.589)$

         b   The fact that 0 is not contained in the confidence interval does not imply that *all* students lie to their mothers. There may be students in the population and even in this sample of 77 who did not lie to their mothers. Even though the mean may not be zero, some of the individual data values may be zero. However, if the mean is nonzero, it does imply that some students tell lies to their mothers.

9.37   **a**   LBO firms: $0.6214 \pm 1.96(0.4) \Rightarrow 0.6214 \pm 0.7840 \Rightarrow (-0.1626, 1.4054)$

             non-LBO firms : $0.9504 \pm 1.96(0.2076) \Rightarrow 0.9504 \pm 0.4069 \Rightarrow (0.5435, 1.3573)$

         b   The non-LBO interval is narrower because it is based on a larger sample size.

9.39   With n = 25, the degrees of freedom is $n-1 = 25-1 = 24$.

From the t-table, the t critical value is 1.71.

The confidence interval is $2.2 \pm 1.71 \left(\dfrac{1.2}{\sqrt{25}} \right) \Rightarrow 2.2 \pm 0.41 \Rightarrow (1.79, 2.61)$.

9.41   The t critical value for a 90% confidence interval when $df = 10 - 1 = 9$ is 1.83. From the given data, $n = 10$, $\Sigma x = 219$, and $\Sigma x^2 = 4949.92$. From the summary statistics, $\bar{x} = \dfrac{219}{10} = 21.9$

$$s^2 = \dfrac{4949.92 - \dfrac{(219)^2}{10}}{9} = \dfrac{4949.92 - 4796.1}{9} = \dfrac{153.82}{9} = 17.09$$

$s = \sqrt{17.09} = 4.134$.

The 90% confidence interval based on this sample data is

$$\bar{x} \pm (\text{t critical}) \dfrac{s}{\sqrt{n}} \Rightarrow 21.9 \pm (1.83) \dfrac{4.134}{\sqrt{10}} \Rightarrow 21.9 \pm 2.39 \Rightarrow (19.51, 24.29).$$

9.43   Summary statistics for the sample are: $n = 5$, $\bar{x} = 17$, $s = 9.03$

The 95% confidence interval is given by

$$\bar{x} \pm (\text{t critical}) \dfrac{s}{\sqrt{n}} \Rightarrow 17 \pm (2.78) \dfrac{9.03}{\sqrt{5}} \Rightarrow 17 \pm 11.23 \Rightarrow (5.77, 28.23).$$

9.45   Since the sample size is small (n = 17), it would be reasonable to use the t confidence interval only if the population distribution is normal (at least approximately). A histogram of the sample data (see figure below) suggests that the normality assumption is not reasonable for these data. In particular, the values 270 and 290 are much larger than the rest of the data and the distribution is skewed to the right. Under the circumstances the use of the t confidence interval for this problem is not reasonable.

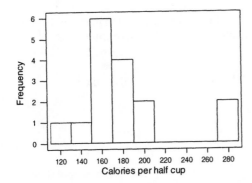

Calories per half cup

9.47   B = 0.1, σ = 0.8

$$n = \left[\frac{1.96\sigma}{B}\right]^2 = \left[\frac{(1.96)(0.8)}{0.1}\right]^2 = (15.68)^2 = 245.86$$

Since a partial observation cannot be taken, n should be *rounded up* to n = 246.

9.49   For 90% confidence level: $n = \left[\frac{(1.645)\sigma}{B}\right]^2$

For 98% confidence level: $n = \left[\frac{(2.33)\sigma}{B}\right]^2$

Supplementary Exercises

9.51   The 99% confidence interval for the mean commuting distance based on this sample is

$$\bar{x} \pm (t\text{ critical})\frac{s}{\sqrt{n}} \Rightarrow 10.9 \pm (2.58)\left(\frac{6.2}{\sqrt{300}}\right) \Rightarrow 10.9 \pm 0.924 \Rightarrow (9.976, 11.824).$$

9.53   **a** $\bar{x} \pm (t\text{ critical})\frac{s}{\sqrt{n}} \Rightarrow 207.8 \pm (2.31)\frac{10.9}{\sqrt{9}} \Rightarrow 207.8 \pm 8.393 \Rightarrow (199.407, 216.193)$

With high confidence (95%), the true mean systolic blood pressure of all anabolic-steroid-using athletes is estimated to be between 199.407 and 216.193 (mm Hg).

b   The validity of the interval in **a** requires that the distribution of systolic blood pressure of anabolic-steroid-using athletes be approximately like a normal distribution.

9.55 $p = \frac{142}{507} = 0.2801$

The 95% confidence interval for the proportion of the entire population that could correctly describe the Bill of Rights as the first ten amendments to the U.S. Constitution is

$$0.2801 \pm 1.96\frac{\sqrt{0.2801(0.7199)}}{507} \Rightarrow 0.2801 \pm 1.96(0.0199) \Rightarrow 0.2801 \pm 0.0391$$

$\Rightarrow (0.2410, 0.3192).$

9.57 The width of the interval discussed in the text is

$$\left(\bar{x}+(1.96)\frac{s}{\sqrt{n}}\right)-\left(\bar{x}-(1.96)\frac{s}{\sqrt{n}}\right)=2(1.96)\frac{s}{\sqrt{n}}=3.92\frac{s}{\sqrt{n}}.$$

The width of the interval suggested in this problem is

$$\left(\bar{x}+(1.75)\frac{s}{\sqrt{n}}\right)-\left(\bar{x}-(2.33)\frac{s}{\sqrt{n}}\right)=(1.75+2.33)\frac{s}{\sqrt{n}}=4.08\frac{s}{\sqrt{n}}.$$

Since this latter interval is wider (less precise) than the one discussed in the text, its use is not recommended.

Chapter 10
Hypothesis Testing Using a Single Sample

Section 10.1

10.1 $\bar{x} = 50$ is not a legitimate hypothesis, because $\bar{x}$ is a statistic, not a population characteristic. Hypotheses are always expressed in terms of a population characteristic, not in terms of a sample statistic.

10.3 If we use the hypothesis $H_o: \mu = 100$ versus $H_a: \mu > 100$, we are taking the position that the welds do not meet specifications, and hence, are not acceptable unless there is substantial evidence to show that the welds are good (i.e. $\mu > 100$). If we use the hypothesis $H_o: \mu = 100$ versus $H_a: \mu < 100$, we initially believe the welds to be acceptable, and hence, they will be declared unacceptable only if there is substantial evidence to show that the welds are faulty. It seems clear that we would choose the first set-up, which places the burden of proof on the welding contractor to show that the welds meet specifications.

10.5 Since the administration has decided to make the change if it can conclude that *more than* 60% of the faculty favor the change, the appropriate hypotheses are

$$H_o: \pi = 0.6 \quad \text{versus} \quad H_a: \pi > 0.6.$$

Then, if H_o is rejected, it can be concluded that more than 60% of the faculty favor a change.

10.7 $H_o: \mu = 170 \quad H_a: \mu < 170$

10.9 Since the manufacturer is interested in detecting values of μ which are less than 40, as well as values of μ which exceed 40, the appropriate hypotheses are: $H_o: \mu = 40 \quad \text{versus} \quad H_a: \mu \neq 40$

Section 10.2

10.11 **a** They failed to reject the null hypothesis, because their conclusion "There is no evidence of increased risk of death due to cancer" for those living in areas with nuclear facilities is precisely what the null hypothesis states.

b   They would be making a type II error since a type I error is failing to reject the null hypothesis when the null is false.

c   Since the null hypothesis is the initially favored hypothesis and is presumed to be the case until it is determined to be false, if we fail to reject the null hypothesis, it is not proven to be true. There is just not sufficient evidence to refute its presumed truth. On the other hand, if the hypothesis test is conducted with

H_o : π is greater than the value for areas without nuclear facilities
H_a : π is less than or equal to the value for areas without nuclear facilities

and the null hypothesis is rejected, then there would be evidence based on data against the presumption of an increased cancer risk associated with living near a nuclear power plant. This is as close as one can come to "proving" the absence of an increased risk using statistical studies.

10.13   **a**   Pizza Hut's decision is consistent with the decision of rejecting H_o.

b   Rejecting H_o when it is true is called a type I error. So if they incorrectly reject H_o, they are making a type I error.

10.15   **a**   A type I error is returning to the supplier a shipment which is not of inferior quality. A type II error is accepting a shipment of inferior quality.

b   The calculator manufacturer would most likely consider a type II error more serious, since they would then end up producing defective calculators.

c   From the supplier's point of view, a type I error would be more serious, because the supplier would end up having lost the profits from the sale of the good printed circuits.

10.17   **a**   The manufacturer claims that the percentage of defective flares is 10%. Certainly one would not object if the proportion of defective flares is less than 10%. Thus, one's primary concern would be if the proportion of defective flares exceeds the value stated by the manufacturer.

b   A type I error entails concluding that the proportion of defective flares exceeds 10% when, in fact, the proportion is 10% or less. The consequence of this decision would be the filing of charges of false advertising against the manufacturer, who is not guilty of such actions. A type II error entails concluding that the proportion of defective flares is 10% when, in reality, the proportion is in excess of 10%. The consequence of this decision is to allow the manufacturer who is guilty of false advertising to continue bilking the consumer.

10.19   **a**   The manufacturer should test the hypotheses
H_o: $\pi = 0.02$ versus H_a: $\pi < 0.02$.

The null hypothesis is implicitly stating that the proportion of defective installations using robots is 0.02 or larger. That is, at least as large as for humans. In other words, they will not undertake to use robots unless it can be shown quite strongly that the defect rate *is less* for robots than for humans.

b   A type I error is changing to robots when in fact they are not superior to humans. A type II error is not changing to robots when in fact they are superior to humans.

c   Since a type I error means substantial loss to the company as well as to the human employees who would become unemployed a small α should be used. Therefore, $\alpha = 0.01$ is preferred.

Section 10.3

10.21 H_o is rejected if P-value $\leq \alpha$. Since $\alpha = 0.05$, H_o should be rejected for the following P-values:

a: 0.001, **b:** 0.021, and **d:** 0.047

10.23   Using Table II from the Appendix we get the following:

a   0.0808

b   0.1762

c   0.0250

d   0.0071

e   0.5675

10.25   1.   Let π represent the proportion of APL patients receiving arsenic who go into remission.

2. $H_o: \pi = 0.15$

3. $H_a: \pi > 0.15$

4. $\alpha = 0.01$

5.   Since $n\pi = 40(0.15) = 6 \geq 5$, and $n(1 - \pi) = 40(0.85) = 34 \geq 5$, the large sample z test may be used.

$$z = \frac{p - 0.15}{\sqrt{\dfrac{0.15(0.85)}{n}}}$$

6. $n = 40$, $p = 0.42$

$$z = \frac{0.42 - 0.15}{\sqrt{\dfrac{0.15(0.85)}{40}}} = \frac{0.27}{0.0565} = 4.782$$

7. P-value = area under the z curve to the right of $4.782 \approx 1 - 1 = 0$

8. Since the P-value is less than α, H_o is rejected. At level of significance 0.01, the data supports the conclusion that the proportion in remission for the arsenic treatment is greater than 0.15.

10.27 1. Let π represent the proportion of U.S. adults who are aware that an investment of $25 a week could result in a sum of over $100,000 over 40 years.

2. $H_o: \pi = 0.50$

3. $H_a: \pi < 0.50$

4. $\alpha = 0.05$.

5. Since $n\pi = 1010(0.5) \geq 5$, and $n(1 - \pi) = 1010(0.5) \geq 5$, the large sample z test may be used.

$$z = \frac{p - 0.5}{\sqrt{\dfrac{0.5(0.5)}{n}}}$$

6. $n = 1010$, $p = 374/1010 = 0.3703$

$$z = \frac{0.3703 - 0.5}{\sqrt{\dfrac{0.5(0.5)}{1010}}} = \frac{-0.1297}{0.01573} = -8.244$$

7. P-value = area under the z curve to the left of $-8.244 \approx 0.0$.

8. Since the P-value is less than α, H_o can be rejected. There is sufficient evidence to conclude that less than half of U.S. adults are aware that an investment of $25 a week could result in a sum of over $100,000 over 40 years.

10.29 **a** 1. Let π represent the proportion of valid signatures on the petition.

2. $H_o: \pi = 0.88$

3. $H_a: \pi < 0.88$

4. $\alpha = 0.01$

5. Since $n\pi = 592(0.88) \geq 5$, and $n(1 - \pi) = 592(0.12) \geq 5$, the large sample z test may be used.

$$z = \frac{p - 0.88}{\sqrt{\dfrac{0.88(0.12)}{n}}}$$

6. $n = 592$, $p = 411/592 = 0.6943$

$$z = \frac{0.6943 - 0.88}{\sqrt{\dfrac{0.88(0.12)}{592}}} = \frac{-0.1857}{0.01336} = -13.9073$$

7. P-value = area under the z curve to the left of -13.9073 ≈ 0.

8. Since the P-value is less than α, H_o is rejected. There is sufficient evidence to conclude that the proportion of valid signatures on the petition is less than 0.88.

b A type-I error would result in this situation if it is concluded that the proportion of valid signatures on the petition is less than 0.88 when in fact it is 0.88 or more. The consequence of such an error would be that a recall election will not be held when it should have been held.

c A type-II error would result in this situation if it is concluded that the proportion of valid signatures on the petition is 0.88 or greater when in fact this proportion is less than 0.88. The consequence would be that a recall election would be held when in fact it should not be held.

d The conclusion to part **a** was to not conduct a recall election. If this conclusion is in error then we would be making a type-I error (rejecting the null hypothesis when it is in fact true).

e Yes, the conclusion of the hypothesis is in agreement with the county clerk's decision to "toss out" the recall petitions.

10.31 1. Let π represent the proportion of job applicants in California that test positive for drug use.

2. H_o: $\pi = 0.10$

3. H_a: $\pi > 0.10$

4. $\alpha = 0.01$ (A value for α was not specified in the problem, so this value was chosen.)

5. Since $n\pi = 600(0.10) = 60 \geq 5$, and $n(1 - \pi) = 600(0.90) = 540 \geq 5$, the large sample z test may be used.

$$z = \frac{p - 0.1}{\sqrt{\frac{0.1(0.9)}{n}}}$$

6. $n = 600$, $x = 73$, $p = \dfrac{73}{600} = 0.1217$

$$z = \frac{0.1217 - 0.1}{\sqrt{\frac{0.1(0.9)}{600}}} = \frac{0.0217}{0.0122} = 1.78$$

7.   P-value = area under the z curve to the right of $1.78 = 1 - 0.9625 = 0.0375$.

8.   Since the P-value exceeds α, H_o is not rejected. At level of significance 0.01, the data does not support the conclusion that the proportion of job applicants in California that test positive for drug use exceeds 0.10.

10.33

1.   Let π represent the proportion of social sciences and humanities majors who have a B average going into college but end up with a GPA below 3.0 at the end of their first year.

2. H_o: $\pi = 0.50$

3. H_a: $\pi > 0.50$

4.   We will compute a P-value for this test. The problem does not specify a value for α but we will use $\alpha = 0.05$ for illustration.

5.   Since $n\pi = 137(0.5) \geq 5$, and $n(1 - \pi) = 137(0.5) \geq 5$, the large sample z test may be used.

$$z = \frac{p - 0.5}{\sqrt{\frac{0.5(0.5)}{n}}}$$

6. $n = 137$, $p = 0.532$

$$z = \frac{0.532 - 0.5}{\sqrt{\frac{0.5(0.5)}{137}}} = \frac{0.032}{0.0427} = 0.7491$$

7.   P-value = area under the z curve to the right of $0.7491 = 0.2269$.

8.   Since the P-value is greater than α, H_o cannot be rejected. At level of significance 0.05, the data do not support the conclusion that a majority of students majoring in

social sciences and humanities who enroll with a HOPE scholarship will lose their scholarship.

10.35 1. Let π denote the proportion of all students in the population sampled that engaged in this form of cheating.

2. $H_o: \pi = 0.20$

3. $H_a: \pi > 0.20$

4. $\alpha = 0.05$ (A value for α was not specified in the problem, so this value was chosen.)

5. Since $n\pi = 480(0.2) \geq 5$, and $n(1 - \pi) = 480(0.8) \geq 5$, the large sample z test may be used.

$$z = \frac{p - 0.20}{\sqrt{\dfrac{0.20(0.80)}{n}}}$$

6. $n = 480$, $x = 124$, $p = \dfrac{124}{480} = 0.258333$

$$z = \frac{0.258333 - 0.20}{\sqrt{\dfrac{0.20(0.80)}{480}}} = \frac{0.058333}{0.018257} = 3.195 \approx 3.20$$

7. P-value = area under the z curve to the right of $3.20 = 1 - 0.9993 = 0.0007$.

8. Since the P-value is less than α, H_o is rejected. There is sufficient evidence in this sample to support the conclusion that the true proportion of students that engage in this form of cheating exceeds 0.20.

10.37 1. Let π denote the proportion of all California lawyers who feel that the ethical standards of most lawyers are high.

2. $H_o: \pi = 0.5$

3. $H_a: \pi < 0.5$

4. $\alpha = 0.05$ (A value for α was not specified in the problem, so this value was chosen.)

5. Since $n\pi = 2700(0.5) \geq 5$, and $n(1 - \pi) = 2700(0.5) \geq 5$, the large sample z test may be used.

$$z = \frac{p - 0.5}{\sqrt{\dfrac{0.5(0.5)}{n}}}$$

6. $n = 2700, x = 1107, \ p = \dfrac{1107}{2700} = 0.41$

$$z = \dfrac{0.41 - 0.5}{\sqrt{\dfrac{0.5(0.5)}{2700}}} = \dfrac{-0.09}{0.009623} = -9.35$$

7.   P-value = area under the z curve to the left of $-9.35 = 0$

8.   Since the P-value is less than α, H_o is rejected. There is strong evidence in this sample to support the conclusion that less than 50% of all California lawyers feel that the ethical standards of most lawyers are high.

Section 10.4

10.39   **a** $z = 3.00$

   b   P-value = 0.0014

   c H_o is rejected if the P-value is less than α. Since the P-value of 0.0014 is less than α, which is 0.05, H_o should be rejected in favor of the conclusion that the machine is overfilling.

10.41   **a**   P-value = area under the 8 d.f. t curve to the right of $2.0 = 0.040$

   b   P-value = area under the 13 d.f. t curve to the right of $3.2 = 0.003$

   c   P-value = area under the 10 d.f. t curve to the left of -2.4
          = area under the 10 d.f. t curve to the right of $2.4 = 0.019$

   d   P-value = area under the 21 d.f. t curve to the left of -4.2
          = area under the 21 d.f. t curve to the right of $4.2 = 0.0002$

   e   P-value = 2(area under the 15 d.f. t curve to the right of 1.6) = 2 (0.065) = 0.13

   f   P-value = 2(area under the 15 d.f. t curve to the right of 1.6) = 2 (0.065) = 0.13

   g   P-value = 2(area under the 15 d.f. t curve to the right of 6.3) = 2(0) = 0

10.43   The P-value for this test is equal to the area under the 14 d.f. t curve to the right of $3.2 = 0.003$.

   a $\alpha = 0.05$, reject H_o

   b $\alpha = 0.01$, reject H_o

   c $\alpha = 0.001$, fail to reject H_o

10.45   **a**   P-value = 2(area under the 12 d.f. t curve to the right of 1.6) = 2(0.068) = 0.136.
Since P-value > α, H_o is not rejected.

       b   P-value = 2(area under the 12 d.f. t curve to the left of -1.6)
            $\cdot$ = 2(area under the 12 d.f. t curve to the right of 1.6) = 2(0.068) = 0.136.
Since P-value > α, H_o is not rejected.

       c   P-value = 2(area under the 24 d.f. t curve to the left of -2.6)
            = 2(area under the 24 d.f. t curve to the right of 2.6) = 2(0.008) = 0.016.
Since P-value > α, H_o is not rejected.

       d   P-value = 2(area under the 24 d.f. t curve to the left of -3.6)
            = 2(area under the 24 d.f. t curve to the right of 3.6) = 2(0.001) = 0.002.
H_o would be rejected for any $\alpha > 0.002$.

10.47   1.   Let μ denote the true average "speaking up" value for Asian men.

       2. H_o: $\mu = 10$

       3. H_a: $\mu < 10$

       4. $\alpha = 0.05$ (A value for α was not specified in the problem, so this value was chosen.)

       5.   Test statistic: $t = \dfrac{\bar{x} - 10}{\dfrac{s}{\sqrt{n}}}$

       6.   Computations: n = 64, $\bar{x} = 8.75$, s = 2.57,

$$t = \frac{8.75 - 10}{\dfrac{2.57}{\sqrt{64}}} = -3.89$$

       7.   P-value = area under the 63 d.f. t curve to the left of -3.89 $\approx$ 0.

       8.   Conclusion: Since the P-value is less than α, H_o is rejected. The data in this sample does support the conclusion that the average "speaking up" score for Asian men is smaller than 10.0.

10.49   1.   Let μ denote the mean rating given by all nonfundamentalists to Christian fundamentalists.

       2. H_o: $\mu = 57$

       3. H_a: $\mu < 57$

4. $\alpha = 0.01$ (A value for α was not specified in the problem. The value 0.01 was chosen for illustration.)

5. Test statistic: $t = \dfrac{\bar{x} - 57}{\dfrac{s}{\sqrt{n}}}$ with d.f. $= 960 - 1 = 959$

6. Computations: $n = 960$, $\bar{x} = 47$, $s = 21$

$$t = \dfrac{47 - 57}{\dfrac{21}{\sqrt{960}}} = \dfrac{-10}{0.6778} = -14.7542$$

7. P-value = area under the 959 d.f. t curve to the left of $-14.7542 \approx 0.0$

8. Since the P-value is smaller than α, H_o is rejected. The sample data do provide convincing evidence that the average rating given by all nonfundamentalists to Christian fundamentalists is below 57.

10.51 1. Let μ denote the true average attention span (in minutes) of teenage Australian boys.

2. $H_o: \mu = 5$

3. $H_a: \mu < 5$

4. $\alpha = 0.01$

5. Test statistic: $t = \dfrac{\bar{x} - 5}{\dfrac{s}{\sqrt{n}}}$ with d.f. $= 50 - 1 = 49$

6. Computations: $n = 50$, $\bar{x} = 4$, $s = 1.4$

$$t = \dfrac{4 - 5}{\dfrac{1.4}{\sqrt{50}}} = \dfrac{-1}{0.198} = -5.0508$$

7. P-value = area under the 49 d.f. t curve to the left of $-5.0508 \approx 0.0$

8. Since the P-value is smaller than α, H_o is rejected. The sample does provide convincing evidence that the average attention span of Australian teenagers is less than 5 minutes.

10.53 a Since the boxplot is nearly symmetric and the normal probability plot is very much like a straight line, it is reasonable to use a t-test to carry out the hypothesis test on μ.

b The median is slightly less than 245 and because of the near symmetry, the mean should be close to 245. Also, because of the large amount of variability in the data, it is quite conceivable that the average calorie content is 240.

c 1. Let μ denote the true average calorie content of this type of frozen dinner.

2. $H_o: \mu = 240$

3. $H_a: \mu \neq 240$

4. $\alpha = 0.05$ (A value for α was not specified in the problem, so this value was chosen.)

5. Test statistic: $t = \dfrac{\bar{x} - 240}{\dfrac{s}{\sqrt{n}}}$ with d.f. $= 12 - 1 = 11$

6. Computations: $n = 12$, $\bar{x} = 244.333$, $s = 12.383$

$$t = \frac{244.333 - 240}{\dfrac{12.383}{\sqrt{12}}} = \frac{4.333}{3.575} = 1.21$$

7. P-value $= 2$(area under the 11 d.f. t curve to the right of 1.21) $\approx 2\,(0.128) = 0.256$

8. Since the P-value exceeds α, H_o is not rejected. The sample evidence does not support the conclusion that the average calorie content differs from 240.

10.55 a 1. Let μ denote the true average activation time to first sprinkler activation using aqueous film forming foam.

2. $H_o: \mu \leq 25$

3. $H_a: \mu > 25$

4. $\alpha = 0.05$

5. $t = \dfrac{\bar{x} - 25}{\dfrac{s}{\sqrt{n}}}$ with d.f. $= 13 - 1 = 12$

6. From the data: $n = 13$, $\bar{x} = 27.92$, $s = 5.62$,

$$t = \frac{27.92 - 25}{\dfrac{5.62}{\sqrt{13}}} = 1.875$$

7. P-value = area under the 12 d.f. t curve to the right of 1.875. From Appendix Table IV, $0.049 > \text{P-value} > 0.041$.

8. Since the P-value is less than α, H_o would be rejected. The conclusion would be that the design specifications have been met.

The assumption being made is that activation time has a normal distribution.

b The P-value reported by Minitab is 0.043, which is close to the value obtained in part **a**. The reported P-value is consistent with the bounds computed in part **a**.

10.57 **a** The z statistic given in this problem is an appropriate test statistic in this setting because:

i The parameter being tested is a population mean.
ii The variance of the population being sampled is assumed known.
iii The sample size is sufficiently large, and hence by the central limit theorem, the distribution of the random variable $\bar{x}$ should be approximately normal.

b A type I error involves concluding that the water being discharged from the power plant has a mean temperature in excess of $150°$ F when, in fact, the mean temperature is not greater than $150°$ F. A type II error is concluding that the mean temperature of water being discharged is $150°$ F or less when, in fact, the mean temperature is in excess of $150°$ F.

c From Appendix Table II, the area to the right of 1.8 is 0.0359. Hence, rejecting H_o when $z > 1.8$ corresponds to an α value of 0.0359.

d If $z > 1.8$, then $\dfrac{\bar{x} - 150}{\dfrac{10}{\sqrt{50}}} > 1.8$, and it follows that $\bar{x} > 150 + 1.8\left(\dfrac{10}{\sqrt{50}}\right) = 152.546$.

In the graph below, the shaded area = P(type II error when $\mu = 153$)

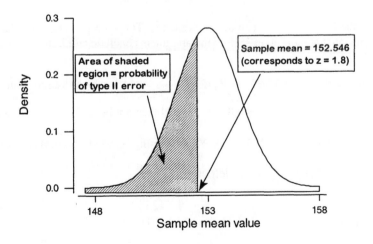

e $\quad \beta$ (when $\mu = 153$) $= P(\overline{x} < 152.546) = P\left[z < \dfrac{152.546 - 153}{\dfrac{10}{\sqrt{50}}}\right] = P(z < -0.32) = 0.3745$

f $\quad \beta$ (when $\mu = 160$) $= P(\overline{x} < 152.546) = P\left[z < \dfrac{152.546 - 160}{\dfrac{10}{\sqrt{50}}}\right] = P(z < -5.27) \approx 0$

g $\quad$ From part **d**, H_o is rejected if $\overline{x} > 152.546$. Since $\overline{x} = 152.4$, H_o is not rejected. Because H_o is not rejected, a type II error might have been made.

10.59 **a** $\quad$ Let π be the true proportion of apartments which prohibit children.

H_o: $\pi = 0.75$

H_a: $\pi > 0.75$

$\alpha = 0.05$

Since $n\pi = 125(0.75) = 93.75 \geq 5$, and $n(1 - \pi) = 125(0.25) = 31.25 \geq 5$, the large sample z test for π may be used.

$$z = \frac{p - 0.75}{\sqrt{\dfrac{0.75(0.25)}{125}}} = \frac{0.816 - 0.75}{0.0387} = 1.71$$

P-value = area under the z curve to the right of $1.71 = 1 - 0.9564 = 0.0436$.

Since the P-value is less than α, H_o is rejected. This 0.05 level test does lead to the conclusion that more than 75% of the apartments exclude children.

b $\quad$ The test with $\alpha = 0.05$ rejects H_o if $\dfrac{p - 0.75}{0.0387} > 1.645$, which is equivalent to

$p > 0.75 + 0.0387(1.645) = 0.8137$. H_o will then not be rejected if $p \leq 0.8137$.

When $\pi = 0.80$ and $n = 125$, $\beta = P$(not rejecting H_o when $\pi = 0.8$) $= P(p \leq 0.8137)$

$=$ area under the z curve to the left of $\dfrac{0.8137 - 0.8}{\sqrt{\dfrac{0.8(0.2)}{125}}}$

$=$ area under the z curve to the left of $0.38 = 0.6480$.

10.61 **a** $\quad d = \dfrac{|\text{alternative value - hypotheszed value}|}{\sigma}$

   i $\quad d = \dfrac{|52-50|}{10} = 0.2$ From Appendix Table V, $\beta \approx 0.85$.

   ii $\quad d = \dfrac{|55-50|}{10} = 0.5$ From Appendix Table V, $\beta \approx 0.55$.

   iii $\quad d = \dfrac{|60-50|}{10} = 1$ From Appendix Table V, $\beta \approx 0.10$.

   iv $\quad d = \dfrac{|70-50|}{10} = 2$ From Appendix Table V, $\beta \approx 0$.

   b $\quad$ As σ increases, d decreases in value. Therefore, the value of β increases.

Supplementary Exercises

10.63 **a** $\quad$ *

   b $\quad$ †

   c $\quad$ none

   d $\quad$ ‡

10.65 Let π represent the true proportion of all firms who offer stock ownership plans to employees because of tax-related benefits.

$H_o: \pi = 0.5$

$H_a: \pi > 0.5$

$\alpha = 0.05$ (A value for α was not specified in the problem, so this value was chosen.)

$$z = \dfrac{p-0.5}{\sqrt{\dfrac{0.5(0.5)}{n}}}$$

$n = 87, x = 54, p = \dfrac{54}{87} = 0.62069$

$$z = \dfrac{0.62069-0.5}{\sqrt{\dfrac{0.5(0.5)}{87}}} = \dfrac{0.12069}{0.053606} = 2.25$$

P-value = area under the z curve to the right of 2.25 = 1 − 0.9878 = 0.0122.

Since the P-value is less than α, H_o is rejected. At level of significance 0.05, the data does support the conclusion that the proportion of firms who offer stock ownership to employees because of tax-related benefits exceeds 0.5.

10.67   **a**   Daily caffeine consumption cannot be a negative value. Since the standard deviation is larger than the mean, this would imply that a sizable portion of a normal curve with this mean and this standard deviation would extend into the negative values on the number line. Therefore, it is not plausible that the population distribution of daily caffeine consumption is normal.

Since the sample size is large (greater than 30) the Central Limit Theorem allows for the conclusion that the distribution of $\overline{x}$ is approximately normal even though the population distribution is not normal. So it is not necessary to assume that the population distribution of daily caffeine consumption is normal to test hypotheses about the value of population mean consumption.

  b   Let μ denote the population mean daily consumption of caffeine.

$H_o : \mu = 200$

$H_a : \mu > 200$

$\alpha = 0.10$

$$t = \frac{\overline{x} - 200}{\dfrac{s}{\sqrt{n}}}$$

$n = 47$, $\overline{x} = 215$, $s = 235$

$$t = \frac{\overline{x} - 200}{\dfrac{s}{\sqrt{n}}} = \frac{215 - 200}{\dfrac{235}{\sqrt{47}}} = 0.44$$

P-value = area under the 46 d.f. t curve to the right of 0.44 = 1 − 0.6700 = 0.33

Since the P-value exceeds the level of significance of 0.10, H_o is not rejected. The data does not support the conclusion that the population mean daily caffeine consumption exceeds 200 mg.

10.69   Let π represent the response rate when the distributor is stigmatized by an eye patch.

$H_o: \pi = 0.40$

$H_a: \pi > 0.40$

$\alpha = 0.05$

The test statistic is: $z = \dfrac{p - 0.40}{\sqrt{\dfrac{(0.40)(0.60)}{n}}}$.

From the data: $n = 200$, $p = \dfrac{109}{200} = 0.545$,

$$z = \dfrac{0.545 - 0.40}{\sqrt{\dfrac{(0.40)(0.60)}{200}}} = 4.19$$

P-value = area under the z curve to the right of $4.19 \approx 1 - 1 = 0$.

Since the P-value is less than the α value of 0.05, H_o is rejected. The data strongly suggests that the response rate does exceed the rate in the past.

10.71 Let μ denote the mean root length of kidney bean plants that are watered with the 50% wastewater solution.

H_o: $\mu = 5.20$

H_a: $\mu > 5.20$

$\alpha = 0.05$

The test statistic is: $t = \dfrac{\bar{x} - 5.20}{\dfrac{s}{\sqrt{n}}}$.

From the sample data, $t = \dfrac{5.46 - 5.20}{\dfrac{0.55}{\sqrt{40}}} = 2.99$.

P-value = area under the 30 d.f. t curve to the right of $2.99 = 1 - .9976 = 0.0024$.

Since the P-value is less than α, the null hypothesis is rejected. It can be concluded that the mean root length is larger for kidney bean plants that are irrigated with the 50% wastewater solution.

10.73   Let μ denote the true average daily revenue subsequent to the price increase.

H_o: $\mu = 50$

H_a: $\mu < 50$

$\alpha = 0.05$

The test statistic is: $t = \dfrac{\bar{x} - 50}{\dfrac{s}{\sqrt{n}}}$ with d.f. = 19.

From the sample: n = 20, $\bar{x} = 47.3$, s = 4.2,

$t = \dfrac{47.3 - 50}{\dfrac{4.2}{\sqrt{20}}} = -2.87$

P-value = area under the 19 d.f. t curve to the left of $-2.87 \approx 0.005$.

Since the P-value is less than α, the null hypothesis is rejected. The data does suggest that the true average daily revenue has decreased from its value prior to the price increase.

10.75   **a**   Let π denote the proportion of voters in a certain state who favor a particular proposed constitutional amendment.

H_o: $\pi = 0.5$

H_a: $\pi > 0.5$

$\alpha = 0.05$

The test statistic is: $z = \dfrac{p - 0.5}{\sqrt{\dfrac{(0.5)(0.5)}{n}}}$.

My sample of random numbers is:

3 1 3 0 9 7 2 5 7 9 0 0 3 4 1 2 7 3 7 9 2 1 5 3 2
8 5 5 7 5 1 3 2 5 1 9 3 3 8 3 8 7 5 5 5 2 4 4 3 4

Let the numbers 0, 1, 2, 3, and 4 represent favoring the proposed constitutional amendment.

n = 50, x = 28, p = 0.56

$$z = \frac{0.56 - 0.5}{\sqrt{\frac{0.5(0.5)}{50}}} = \frac{0.06}{0.070711} = 0.85$$

P-value = area under the z curve to the right of 0.85 = $1 - 0.802 = 0.198$.

Since the P-value of 0.198 exceeds the alpha of 0.05, H_o is not rejected. The data does not support the conclusion that the true proportion of voters who favor a particular proposed constitutional amendment exceeds 0.5.

b   About $0.05(100) = 5$ times.

c   In this sample, let the numbers 0, 1, 2, 3, 4, and 5 represent favoring the constitutional amendment.

9 6 6 4 2 0 9 5 1 9 4 2 6 4 2 6 5 6 1 9 2 3 7 8 8
0 1 5 0 1 5 0 5 0 4 6 9 5 7 3 7 0 0 6 4 3 1 7 1 9

$n = 50, \; x = 31, \; p = 0.62, \quad z = \frac{0.62 - 0.5}{\sqrt{\frac{0.5(0.5)}{50}}} = \frac{0.12}{0.070711} = 1.697$

P-value = area under the z curve to the right of 1.697 = $1 - 0.9552 = 0.0448$.

Since the P-value of 0.0446 is less than the alpha of 0.05, H_o is rejected. The data does support the conclusion that the true proportion of voters who favor a particular proposed constitutional amendment exceeds 0.5. You would expect (and hope) that you would reject a false null hypothesis more often than you would reject a true null hypothesis.

Chapter 11
Comparing Two Populations or Treatments

Section 11.1

11.1 $\mu_{\bar{x}_1-\bar{x}_2} = \mu_1 - \mu_2 = 30 - 25 = 5$

$$\sigma_{\bar{x}_1-\bar{x}_2} = \sqrt{\frac{\sigma_1^2}{n_1} + \frac{\sigma_2^2}{n_2}} = \sqrt{\frac{(2)^2}{40} + \frac{(3)^2}{50}} = \sqrt{\frac{4}{40} + \frac{9}{50}} = \sqrt{0.28} = 0.529$$

Since both n_1 and n_2 are large, the sampling distribution of $\bar{x}_1 - \bar{x}_2$ is approximately normal. It is centered at 5 and the standard deviation is 0.529.

11.3   **a** $H_o: \mu_1 - \mu_2 = 10$ $H_a: \mu_1 - \mu_2 > 10$

b $H_o: \mu_1 - \mu_2 = -10$ $H_a: \mu_1 - \mu_2 < -10$

11.5   **a**   Small prey:

Let μ_1 be the mean amount (mg) of venom injected by the inexperienced snakes and μ_2 the mean amount of venom injected by the experienced snakes when the prey is a small prey.

$H_o: \mu_1 - \mu_2 = 0$ $H_a: \mu_1 - \mu_2 \neq 0$

$\alpha = 0.05$ (A value for α is not specified in the problem. We use $\alpha=0.05$ for illustration.)

Test statistic: $t = \dfrac{(\bar{x}_1 - \bar{x}_2) - 0}{\sqrt{\dfrac{s_1^2}{n_1} + \dfrac{s_2^2}{n_2}}}$

Assumptions: The population distributions are (at least approximately) normal and the two samples are independently selected random samples.

$n_1 = 7$, $\bar{x}_1 = 3.1$, $s_1 = 1.0$, $n_2 = 7$, $\bar{x}_2 = 2.6$, $s_2 = 0.3$

$$t = \frac{(3.1-2.6)-0}{\sqrt{\frac{(1.0)^2}{7} + \frac{(0.3)^2}{7}}} = \frac{0.5}{0.3946} = 1.2670$$

$$df = \frac{\left(\frac{s_1^2}{n_1} + \frac{s_2^2}{n_2}\right)^2}{\frac{1}{n_1-1}\left(\frac{s_1^2}{n_1}\right)^2 + \frac{1}{n_2-1}\left(\frac{s_2^2}{n_2}\right)^2} = \frac{(0.1428+0.0128)^2}{\frac{(0.1428)^2}{6} + \frac{(0.0128)^2}{6}} = 7.0713$$

So $df = 7$ (rounded down to an integer)

P-value = 2 times the area under the 7 df t curve to the right of $1.2670 \approx 0.2456$.

Since the P-value is greater than α, the null hypothesis cannot be rejected. At level of significance 0.05 (or even 0.10), the data do not indicate that there is a difference in the amount of venom injected between inexperienced snakes and experienced snakes when the prey is a small prey.

b   Medium prey:

Let μ_1 be the mean amount (mg) of venom injected by the inexperienced snakes and μ_2 the mean amount of venom injected by the experienced snakes for medium prey.

$H_o: \mu_1 - \mu_2 = 0$ $H_a: \mu_1 - \mu_2 \neq 0$

$\alpha = 0.05$ (A value for α is not specified in the problem. We use $\alpha=0.05$ for illustration.)

Test statistic: $t = \dfrac{(\bar{x}_1 - \bar{x}_2)-0}{\sqrt{\frac{s_1^2}{n_1} + \frac{s_2^2}{n_2}}}$

Assumptions: The population distributions are (at least approximately) normal and the two samples are independently selected random samples.

$n_1 = 7$, $\bar{x}_1 = 3.4$, $s_1 = 0.4$, $n_2 = 7$, $\bar{x}_2 = 2.9$, $s_2 = 0.6$

$$t = \frac{(3.4-2.9)-0}{\sqrt{\frac{(0.4)^2}{7} + \frac{(0.6)^2}{7}}} = \frac{0.5}{0.2725} = 1.8344$$

$$df = \frac{\left(\frac{s_1^2}{n_1} + \frac{s_2^2}{n_2}\right)^2}{\frac{1}{n_1-1}\left(\frac{s_1^2}{n_1}\right)^2 + \frac{1}{n_2-1}\left(\frac{s_2^2}{n_2}\right)^2} = \frac{(0.0228+0.0514)^2}{\frac{(0.0228)^2}{6} + \frac{(0.0514)^2}{6}} = 10.45$$

So $df = 10$ (rounded down to an integer)

P-value = 2 times the area under the 10 df t curve to the right of $1.8344 \approx 0.0964$.

Since the P-value is greater than α, the null hypothesis cannot be rejected. At level of significance 0.05, the data do not indicate that there is a difference in the amount of venom injected between inexperienced snakes and experienced snakes for medium prey.

c Large prey:

Let μ_1 be the mean amount (mg) of venom injected by the inexperienced snakes and μ_2 the mean amount of venom injected by the experienced snakes for large prey.

$H_o: \mu_1 - \mu_2 = 0$

$H_a: \mu_1 - \mu_2 \neq 0$

$\alpha = 0.05$ (A value for α is not specified in the problem. We use $\alpha = 0.05$ for illustration.)

Test statistic: $t = \dfrac{(\bar{x}_1 - \bar{x}_2) - 0}{\sqrt{\dfrac{s_1^2}{n_1} + \dfrac{s_2^2}{n_2}}}$

Assumptions: The population distributions are (at least approximately) normal and the two samples are independently selected random samples.

$n_1 = 7$, $\bar{x}_1 = 1.8$, $s_1 = 0.3$, $n_2 = 7$, $\bar{x}_2 = 4.7$, $s_2 = 0.3$

$$t = \frac{(1.8-4.7)-0}{\sqrt{\frac{(0.3)^2}{7} + \frac{(0.3)^2}{7}}} = \frac{-2.9}{0.1603} = -18.0846$$

$$df = \frac{\left(\frac{s_1^2}{n_1} + \frac{s_2^2}{n_2}\right)^2}{\frac{1}{n_1-1}\left(\frac{s_1^2}{n_1}\right)^2 + \frac{1}{n_2-1}\left(\frac{s_2^2}{n_2}\right)^2} = \frac{(0.0128+0.0128)^2}{\frac{(0.0128)^2}{6} + \frac{(0.0128)^2}{6}} = 12.0.$$ So $df = 12$

P-value = 2 times the area under the 12 df t curve to the left of -18.0846 ≈ 0.0000. Since the P-value is smaller than α, the null hypothesis is rejected. The data provide strong evidence that there is a difference in the amount of venom injected between inexperienced snakes and experienced snakes for large prey.

11.7 **a**   Let μ_1 denote the true mean level of testosterone for male trial lawyers and μ_2 the true mean level of testosterone for male nontrial lawyers. We wish to test the null hypothesis $H_o: \mu_1 - \mu_2 = 0$ against the alternative hypothesis $H_a: \mu_1 - \mu_2 \neq 0$. The t statistic for this test is reported to be 3.75 and the degrees of freedom are 64. The P-value is twice the area under the 64 df t curve to the right of 3.75. This is equal to 0.0004 (the report states that the P-value is < 0.001 but doesn't report the actual P-value). Hence the data do provide strong evidence to conclude that the mean testosterone levels for male trial lawyers and nontrial lawyers are different.

b   Let μ_1 denote the true mean level of testosterone for female trial lawyers and μ_2 the true mean level of testosterone for female nontrial lawyers. We wish to test the null hypothesis $H_o: \mu_1 - \mu_2 = 0$ against the alternative hypothesis $H_a: \mu_1 - \mu_2 \neq 0$. The t-statistic for this test is reported to be 2.26 and the degrees of freedom are 29. The P-value is twice the area under the 29 df t curve to the right of 2.26. This is equal to 0.0316 (the report states that the P-value is < 0.05 but doesn't report the actual P-value). Hence the data do provide sufficient evidence to conclude that the mean testosterone levels for female trial lawyers and female nontrial lawyers are different.

c   There is not enough information to carry out a test to determine whether there is a significant difference in the mean testosterone levels of male and female trial lawyers. To carry out such a test we need the sample means and sample standard deviations for the 35 male trial lawyers and the 13 female trial lawyers.

11.9 **a**   Let μ_1 be the true mean "appropriateness" score assigned to wearing a hat in a class by the population of students and μ_2 be the corresponding score for faculty.

$H_o: \mu_1 - \mu_2 = 0$ $H_a: \mu_1 - \mu_2 \neq 0$

$\alpha = 0.05$ (A value for α is not specified in the problem. We use α=0.05 for illustration.)

Test statistic: $t = \dfrac{(\overline{x}_1 - \overline{x}_2) - 0}{\sqrt{\dfrac{s_1^2}{n_1} + \dfrac{s_2^2}{n_2}}}$

Assumptions: The sample sizes for the two groups are large (say, greater than 30 for each) and the two samples are independently selected random samples.

$n_1 = 173$, $\overline{x}_1 = 2.80$, $s_1 = 1.0$, $n_2 = 98$, $\overline{x}_2 = 3.63$, $s_2 = 1.0$

$$t = \frac{(2.80-3.63)-0}{\sqrt{\dfrac{(1.0)^2}{173}+\dfrac{(1.0)^2}{98}}} = \frac{-0.83}{0.1264} = -6.5649$$

$$df = \frac{\left(\dfrac{s_1^2}{n_1}+\dfrac{s_2^2}{n_2}\right)^2}{\dfrac{1}{n_1-1}\left(\dfrac{s_1^2}{n_1}\right)^2+\dfrac{1}{n_2-1}\left(\dfrac{s_2^2}{n_2}\right)^2} = \frac{(0.00578+0.01020)^2}{\dfrac{(0.00578)^2}{172}+\dfrac{(0.01020)^2}{97}} = 201.5$$

So $df = 201$ (rounded down to an integer)

P-value = 2 times the area under the 201 df t curve to the left of $-6.5649 \approx 0.0000$.

Since the P-value is much less than α, the null hypothesis of no difference is rejected. The data do provide very strong evidence to indicate that there is a difference in the mean appropriateness scores between students and faculty for wearing hats in the class room. The mean appropriateness score for students is significantly smaller than that for faculty.

b   Let μ_1 be the true mean "appropriateness" score assigned to addressing an instructor by his or her first name by the population of students and μ_2 be the corresponding score for faculty.

$H_o: \mu_1 - \mu_2 = 0 \quad H_a: \mu_1 - \mu_2 > 0$

$\alpha = 0.05$ (A value for α is not specified in the problem. We use $\alpha = 0.05$ for illustration.)

Test statistic: $t = \dfrac{(\bar{x}_1 - \bar{x}_2)-0}{\sqrt{\dfrac{s_1^2}{n_1}+\dfrac{s_2^2}{n_2}}}$

Assumptions: The sample sizes for the two groups are large (say, greater than 30 for each) and the two samples are independently selected random samples.

$n_1 = 173,\ \bar{x}_1 = 2.90,\ s_1 = 1.0,\ n_2 = 98,\ \bar{x}_2 = 2.11,\ s_2 = 1.0$

$$t = \frac{(2.90-2.11)-0}{\sqrt{\dfrac{(1.0)^2}{173}+\dfrac{(1.0)^2}{98}}} = \frac{0.79}{0.1264} = 6.2485$$

$$df = \frac{\left(\dfrac{s_1^2}{n_1} + \dfrac{s_2^2}{n_2}\right)^2}{\dfrac{1}{n_1-1}\left(\dfrac{s_1^2}{n_1}\right)^2 + \dfrac{1}{n_2-1}\left(\dfrac{s_2^2}{n_2}\right)^2} = \frac{(0.00578 + 0.01020)^2}{\dfrac{(0.00578)^2}{172} + \dfrac{(0.01020)^2}{97}} = 201.5$$

So $df = 201$ (rounded down to an integer)

P-value = the area under the 201 df t curve to the right of 6.2485 ≈ 0.0000.
Since the P-value is much less than α, the null hypothesis of no difference is rejected. The data do provide very strong evidence to indicate that the mean appropriateness score for addressing the instructor by his or her first name is higher for students than for faculty.

c   Let μ_1 be the true mean "appropriateness" score assigned to talking on a cell phone during class by the population of students and μ_2 be the corresponding score for faculty.

$H_o: \mu_1 - \mu_2 = 0$ $H_a: \mu_1 - \mu_2 \neq 0$

$\alpha = 0.05$ (A value for α is not specified in the problem. We use $\alpha=0.05$ for illustration.)

Test statistic: $t = \dfrac{(\overline{x}_1 - \overline{x}_2) - 0}{\sqrt{\dfrac{s_1^2}{n_1} + \dfrac{s_2^2}{n_2}}}$

Assumptions: The sample sizes for the two groups are large (say, greater than 30 for each) and the two samples are independently selected random samples.

$n_1 = 173$, $\overline{x}_1 = 1.11$, $s_1 = 1.0$, $n_2 = 98$, $\overline{x}_2 = 1.10$, $s_2 = 1.0$

$$t = \frac{(1.11 - 1.10) - 0}{\sqrt{\dfrac{(1.0)^2}{173} + \dfrac{(1.0)^2}{98}}} = \frac{-0.01}{0.1264} = 0.0791$$

$$df = \frac{\left(\dfrac{s_1^2}{n_1} + \dfrac{s_2^2}{n_2}\right)^2}{\dfrac{1}{n_1-1}\left(\dfrac{s_1^2}{n_1}\right)^2 + \dfrac{1}{n_2-1}\left(\dfrac{s_2^2}{n_2}\right)^2} = \frac{(0.00578 + 0.01020)^2}{\dfrac{(0.00578)^2}{172} + \dfrac{(0.01020)^2}{97}} = 201.5$$

So $df = 201$ (rounded down to an integer)

P-value = 2 times the area under the 201 df t curve to the right of 0.0791 ≈ 0.9370.

Since the P-value is not less than α, the null hypothesis of no difference cannot be rejected. The data do not provide evidence to indicate that there is a difference in the mean appropriateness scores between students and faculty for talking on cell phones in class. The result does not imply that students and faculty consider it acceptable to talk on a cell phone during class. It simply says that data do not provide enough evidence to claim a difference exists.

11.11   a   Let μ_1 be the true mean stream gradient (%) for the population of sites with tailed frogs and μ_2 be the corresponding mean for sites without tailed frogs.

$H_o: \mu_1 - \mu_2 = 0 \quad H_a: \mu_1 - \mu_2 \neq 0$

$\alpha = 0.01$

Test statistic: $\quad t = \dfrac{(\bar{x}_1 - \bar{x}_2) - 0}{\sqrt{\dfrac{s_1^2}{n_1} + \dfrac{s_2^2}{n_2}}}$

Assumptions: The distribution of stream gradients is approximately normal for both types of sites and the two samples are independently selected random samples.

$n_1 = 18$, $\bar{x}_1 = 9.1$, $s_1 = 6.0$, $n_2 = 31$, $\bar{x}_2 = 5.9$, $s_2 = 6.29$

$$t = \frac{(9.1 - 5.9) - 0}{\sqrt{\dfrac{(6.00)^2}{18} + \dfrac{(6.29)^2}{31}}} = \frac{3.2}{1.8100} = 1.7679$$

$$df = \frac{\left(\dfrac{s_1^2}{n_1} + \dfrac{s_2^2}{n_2}\right)^2}{\dfrac{1}{n_1 - 1}\left(\dfrac{s_1^2}{n_1}\right)^2 + \dfrac{1}{n_2 - 1}\left(\dfrac{s_2^2}{n_2}\right)^2} = \frac{(2.00 + 1.2763)^2}{\dfrac{(2.00)^2}{17} + \dfrac{(1.2763)^2}{30}} = 37.07$$

So $df = 37$ (rounded down to an integer)

P-value = 2 times the area under the 37 df t curve to the right of 1.7679 ≈ 0.0853.

Since the P-value is greater than α, the null hypothesis of no difference cannot be rejected. The data do not provide sufficient evidence to suggest a difference between the mean stream gradients for sites with tailed frogs and sites without tailed frogs.

b   Let μ_1 be the true mean water temperature for sites with tailed frogs and μ_2 be the corresponding mean for sites without tailed frogs.

$H_o: \mu_1 - \mu_2 = 0 \quad H_a: \mu_1 - \mu_2 \neq 0$

$\alpha = 0.01$

Test statistic: $t = \dfrac{(\bar{x}_1 - \bar{x}_2) - 0}{\sqrt{\dfrac{s_1^2}{n_1} + \dfrac{s_2^2}{n_2}}}$

Assumptions: The distribution of stream gradients is approximately normal for both types of sites and the two samples are independently selected random samples.

$n_1 = 18$, $\bar{x}_1 = 12.2$, $s_1 = 1.71$, $n_2 = 31$, $\bar{x}_2 = 12.8$, $s_2 = 1.33$

$$t = \frac{(12.2 - 12.8) - 0}{\sqrt{\dfrac{(1.71)^2}{18} + \dfrac{(1.33)^2}{31}}} = \frac{-0.6}{0.4685} = -1.2806$$

$$df = \frac{\left(\dfrac{s_1^2}{n_1} + \dfrac{s_2^2}{n_2}\right)^2}{\dfrac{1}{n_1 - 1}\left(\dfrac{s_1^2}{n_1}\right)^2 + \dfrac{1}{n_2 - 1}\left(\dfrac{s_2^2}{n_2}\right)^2} = \frac{(0.1625 + 0.0571)^2}{\dfrac{(0.1625)^2}{17} + \dfrac{(0.0571)^2}{30}} = 29.01$$

So $df = 29$ (rounded down to an integer)

P-value = 2 times the area under the 29 df t curve to the left of $-1.2806 \approx 0.2105$.

Since the P-value is greater than α, the null hypothesis of no difference cannot be rejected. The data do not provide sufficient evidence to conclude that the mean water temperatures for the two types of sites (with and without tailed frogs) are different.

c Let μ_1 be the true mean stream depth for the population of sites with tailed frogs and μ_2 be the corresponding mean for sites without tailed frogs.

$H_o: \mu_1 - \mu_2 = 0$ $\quad$ $H_a: \mu_1 - \mu_2 \neq 0$

$\alpha = 0.01$

Test statistic: $t = \dfrac{(\bar{x}_1 - \bar{x}_2) - 0}{\sqrt{\dfrac{s_1^2}{n_1} + \dfrac{s_2^2}{n_2}}}$

Assumptions: The sample sizes for each group is large (greater than or equal to 30) and the two samples are independently selected random samples. Since the sample sizes for the two samples are 82 and 267 respectively, it is quite reasonable to use the independent samples t test for comparing the mean depths for the two types of sites.

$n_1 = 82$, $\bar{x}_1 = 5.32$, $s_1 = 2.27$, $n_2 = 267$, $\bar{x}_2 = 8.46$, $s_2 = 5.95$

$$t = \frac{(5.32 - 8.46) - 0}{\sqrt{\dfrac{(2.27)^2}{82} + \dfrac{(5.95)^2}{267}}} = \frac{-3.14}{0.4421} = -7.1028$$

$$df = \frac{\left(\dfrac{s_1^2}{n_1} + \dfrac{s_2^2}{n_2}\right)^2}{\dfrac{1}{n_1 - 1}\left(\dfrac{s_1^2}{n_1}\right)^2 + \dfrac{1}{n_2 - 1}\left(\dfrac{s_2^2}{n_2}\right)^2} = \frac{(0.06284 + 0.13259)^2}{\dfrac{(0.06284)^2}{81} + \dfrac{(0.13259)^2}{266}} = 332.6$$

So $df = 332$ (rounded down to an integer)

P-value = 2 times the area under the 332 df t curve to the left of $-7.1028 \approx 0.0000$.

Since the P-value is much smaller than α, the null hypothesis of no difference is rejected. The data provide very strong evidence to conclude that there indeed is a difference between the mean stream depths of sites with tailed frogs and sites without tailed frogs.

11.13   **a**   Let μ denote the mean salary (in Canadian dollars) for the population of female MBA graduates of this Canadian business school.

H_o: $\mu = 100,000$ H_a: $\mu > 100,000$

A value for α was not specified in the problem. We will compute the P-value.

Test statistic: $t = \dfrac{\bar{x} - 100,000}{\dfrac{s}{\sqrt{n}}}$ with d.f. $= 233 - 1 = 232$

Computations: n = 233, $\bar{x} = 105,156$, s = 98,525

$$t = \frac{105,156 - 100,000}{\dfrac{98,525}{\sqrt{233}}} = \frac{5156.0}{6454.587} = 0.7988$$

P-value = area under the 232 d.f. t curve to the right of $0.7988 \approx 0.2126$

For significance levels greater than 0.2126 we can conclude that the mean salary of female MBA graduates from this business school is above 100,000 dollars.

b   Let μ_1 be the true mean salary for female MBA graduates from this business school and μ_2 be the mean for male MBA graduates.

H_o: $\mu_1 - \mu_2 = 0$ H_a: $\mu_1 - \mu_2 < 0$

$\alpha = 0.01$ (a value for α is not specified in this problem. We will use $\alpha = 0.01$ for illustration.)

Test statistic: $t = \dfrac{(\bar{x}_1 - \bar{x}_2) - 0}{\sqrt{\dfrac{s_1^2}{n_1} + \dfrac{s_2^2}{n_2}}}$

Assumptions: The sample sizes for each group is large (greater than or equal to 30) and the two samples are independently selected random samples.

$n_1 = 233$, $\bar{x}_1 = 105{,}156$, $s_1 = 98{,}525$, $n_2 = 258$, $\bar{x}_2 = 133{,}442$, $s_2 = 131{,}090$

$t = \dfrac{(105{,}156 - 133{,}442) - 0}{\sqrt{\dfrac{(98{,}525)^2}{233} + \dfrac{(131{,}090)^2}{258}}} = \dfrac{-28{,}286}{10405.22} = -2.718$

$df = \dfrac{\left(\dfrac{s_1^2}{n_1} + \dfrac{s_2^2}{n_2}\right)^2}{\dfrac{1}{n_1 - 1}\left(\dfrac{s_1^2}{n_1}\right)^2 + \dfrac{1}{n_2 - 1}\left(\dfrac{s_2^2}{n_2}\right)^2} = \dfrac{(41{,}661{,}697.96 + 66{,}606{,}930.62)^2}{\dfrac{(41{,}661{,}697.96)^2}{232} + \dfrac{(66{,}606{,}930.62)^2}{257}} = 473.7$

So $df = 473$ (rounded down to an integer)

P-value = the area under the 473 df t curve to the left of $-2.718 \approx 0.0034$.

Since the P-value is much smaller than α, the null hypothesis of no difference is rejected. The data provide very strong evidence to conclude that the mean salary for female MBA graduates from this business school is lower than that for the male MBA graduates.

11.15 Let μ_1 denote the true mean approval rating for male players and μ_2 the true mean approval rating for female players.

H_o: $\mu_1 - \mu_2 = 0$ H_a: $\mu_1 - \mu_2 > 0$

$\alpha = 0.01$

Test statistic: $t = \dfrac{(\bar{x}_1 - \bar{x}_2) - 0}{\sqrt{\dfrac{s_1^2}{n_1} + \dfrac{s_2^2}{n_2}}}$

Assumptions: The sample sizes for each group is large (greater than or equal to 30) and the two samples are independently selected random samples.

$n_1 = 56$, $\bar{x}_1 = 2.76$, $s_1 = 0.44$, $n_2 = 67$, $\bar{x}_2 = 2.02$, $s_2 = 0.41$

$$t = \frac{(2.76 - 2.02) - 0}{\sqrt{\frac{(0.44)^2}{56} + \frac{(0.41)^2}{67}}} = \frac{0.74}{0.0772} = 9.58$$

$$df = \frac{\left(\frac{s_1^2}{n_1} + \frac{s_2^2}{n_2}\right)^2}{\frac{1}{n_1-1}\left(\frac{s_1^2}{n_1}\right)^2 + \frac{1}{n_2-1}\left(\frac{s_2^2}{n_2}\right)^2} = \frac{(0.003457 + 0.002509)^2}{\frac{(0.003457)^2}{55} + \frac{(0.002509)^2}{66}} = 113.8$$

So $df = 113$ (rounded down to an integer)

P-value = area under the 113 df t curve to the right of $9.58 \approx 1 - 1 = 0$.

Since the P-value is less than α, the null hypothesis is rejected. At level of significance 0.05, the data supports the conclusion that the mean approval rating is higher for males than for females.

11.17 **a**   Let μ_1 denote the true mean hardness for chicken chilled 0 hours before cooking and μ_2 the true mean hardness for chicken chilled 2 hours before cooking.

$H_o: \mu_1 - \mu_2 = 0$ $H_a: \mu_1 - \mu_2 \neq 0$

$\alpha = 0.05$

Test statistic: $t = \dfrac{(\bar{x}_1 - \bar{x}_2) - 0}{\sqrt{\dfrac{s_1^2}{n_1} + \dfrac{s_2^2}{n_2}}}$

Assumptions: The sample sizes for each group is large (greater than or equal to 30) and the two samples are independently selected random samples.

$n_1 = 36$, $\bar{x}_1 = 7.52$, $s_1 = 0.96$, $n_2 = 36$, $\bar{x}_2 = 6.55$, $s_2 = 1.74$

$$t = \frac{(7.52 - 6.55) - 0}{\sqrt{\frac{(0.96)^2}{36} + \frac{(1.74)^2}{36}}} = \frac{0.97}{0.33121} = 2.93$$

$$df = \frac{\left(\frac{s_1^2}{n_1} + \frac{s_2^2}{n_2}\right)^2}{\frac{1}{n_1-1}\left(\frac{s_1^2}{n_1}\right)^2 + \frac{1}{n_2-1}\left(\frac{s_2^2}{n_2}\right)^2} = \frac{(0.0256 + 0.0841)^2}{\frac{(0.0256)^2}{35} + \frac{(0.0841)^2}{35}} = 54.5$$

So $df = 54$ (rounded down to an integer)

P-value = 2(area under the 54 df t curve to the right of 2.93) = $2(1 - 0.9975)$ = $2(0.00249) = 0.00498$.

Since the P-value is less than α, the null hypothesis is rejected. At level of significance 0.05, there is sufficient evidence to conclude that there is a difference in mean hardness of chicken chilled 0 hours before cooking and chicken chilled 2 hours before cooking.

b Let μ_1 denote the true mean hardness for chicken chilled 8 hours before cooking and μ_2 the true mean hardness for chicken chilled 24 hours before cooking.

$H_o: \mu_1 - \mu_2 = 0$ $H_a: \mu_1 - \mu_2 \neq 0$

$\alpha = 0.05$

Test statistic: $t = \dfrac{(\bar{x}_1 - \bar{x}_2) - 0}{\sqrt{\dfrac{s_1^2}{n_1} + \dfrac{s_2^2}{n_2}}}$

Assumptions: The sample sizes for each group is large (greater than or equal to 30) and the two samples are independently selected random samples.

$n_1 = 36$, $\bar{x}_1 = 5.70$, $s_1 = 1.32$, $n_2 = 36$, $\bar{x}_2 = 5.65$, $s_2 = 1.50$

$$t = \frac{(5.70 - 5.65) - 0}{\sqrt{\dfrac{(1.32)^2}{36} + \dfrac{(1.50)^2}{36}}} = \frac{0.05}{0.333017} = 0.15$$

$$df = \frac{\left(\dfrac{s_1^2}{n_1} + \dfrac{s_2^2}{n_2}\right)^2}{\dfrac{1}{n_1 - 1}\left(\dfrac{s_1^2}{n_1}\right)^2 + \dfrac{1}{n_2 - 1}\left(\dfrac{s_2^2}{n_2}\right)^2} = \frac{(0.0484 + 0.0625)^2}{\dfrac{(0.0484)^2}{35} + \dfrac{(0.0625)^2}{35}} = 68.9$$

So $df = 68$ (rounded down to an integer)

P-value = 2(area under the 68 df t curve to the right of 0.15) = $2(1 - 0.5595)$ = $2(0.44055) = 0.8811$.

Since the P-value exceeds α, the null hypothesis is not rejected. At level of significance 0.05, there is not sufficient evidence to conclude that there is a difference in mean hardness of chicken chilled 8 hours before cooking and chicken chilled 24 hours before cooking.

c Let μ_1 denote the true mean hardness for chicken chilled 2 hours before cooking and μ_2 the true mean hardness for chicken chilled 8 hours before cooking.

$n_1 = 36$, $\bar{x}_1 = 6.55$, $s_1 = 1.74$, $n_2 = 36$, $\bar{x}_2 = 5.70$, $s_2 = 1.32$

$$(6.55 - 5.70) \pm 1.669 \sqrt{\frac{(1.74)^2}{36} + \frac{(1.32)^2}{36}}$$

$$\Rightarrow .85 \pm 1.669(.364005) \Rightarrow .85 \pm .6075 \Rightarrow (.242, 1.458)$$

Based on this sample, we believe that the mean hardness for chicken chilled for 2 hours before cooking is larger than the mean hardness for chicken chilled 8 hours before cooking. The difference may be as small as 0.242, or may be as large as 1.458.

11.19 Let μ_1 denote the true mean alkalinity for upstream locations and μ_2 the true mean alkalinity for downstream locations.

$H_o: \mu_1 - \mu_2 = 0 \quad H_a: \mu_1 - \mu_2 < 0$

$\alpha = 0.05$

Test statistic: $\quad t = \dfrac{(\bar{x}_1 - \bar{x}_2) - 0}{\sqrt{\dfrac{s_1^2}{n_1} + \dfrac{s_2^2}{n_2}}}$

Assumptions: The distribution of alkalinity is approximately normal for both types of sites (upstream and downstream) and the two samples are independently selected random samples.

$n_1 = 24$, $\bar{x}_1 = 75.9$, $s_1 = 1.83$, $n_2 = 24$, $\bar{x}_2 = 183.6$, $s_2 = 1.70$

$$t = \frac{(75.9 - 183.6) - 0}{\sqrt{\dfrac{(1.83)^2}{24} + \dfrac{(1.70)^2}{24}}} = \frac{-107.7}{0.50986} = -211.22$$

$$df = \frac{\left(\dfrac{s_1^2}{n_1} + \dfrac{s_2^2}{n_2}\right)^2}{\dfrac{1}{n_1 - 1}\left(\dfrac{s_1^2}{n_1}\right)^2 + \dfrac{1}{n_2 - 1}\left(\dfrac{s_2^2}{n_2}\right)^2} = \frac{(0.1395 + 0.1204)^2}{\dfrac{(0.1395)^2}{23} + \dfrac{(0.1204)^2}{23}} = 45.75$$

So $df = 45$ (rounded down to an integer)

P-value = area under the 45 df t curve to the left of -211.22 is practically 0.

Since the P-value is less than α, the null hypothesis is rejected. The data supports the conclusion that the true mean alkalinity score for downstream sites is higher than that for upstream sites.

11.20 Let μ_1 denote the true mean cholesterol level of people who have attempted suicide and μ_2 the true mean cholesterol level of people who have not attempted suicide.

$H_o: \mu_1 - \mu_2 = 0 \quad H_a: \mu_1 - \mu_2 < 0$

$\alpha = 0.05$

Test statistic: $t = \dfrac{(\bar{x}_1 - \bar{x}_2) - 0}{\sqrt{\dfrac{s_1^2}{n_1} + \dfrac{s_2^2}{n_2}}}$

Assumptions: The sample size for each group is large (greater than or equal to 30) and the two samples are independently selected random samples.

$n_1 = 331, \bar{x}_1 = 198, s_1 = 20, n_2 = 331, \bar{x}_2 = 217, s_2 = 24$

$t = \dfrac{(198 - 217)}{\sqrt{\dfrac{(20)^2}{331} + \dfrac{(24)^2}{331}}} = \dfrac{-19}{1.717161} = -11.06$

$df = \dfrac{\left(\dfrac{s_1^2}{n_1} + \dfrac{s_2^2}{n_2}\right)^2}{\dfrac{1}{n_1 - 1}\left(\dfrac{s_1^2}{n_1}\right)^2 + \dfrac{1}{n_2 - 1}\left(\dfrac{s_2^2}{n_2}\right)^2} = \dfrac{(1.2085 + 1.7402)^2}{\dfrac{(1.2085)^2}{330} + \dfrac{(1.7402)^2}{330}} = 639.2$

So $df = 639$ (rounded down to an integer)

P-value = area under the 639 df t curve to the left of $-11.06 \approx 0$.

Since the P-value is less than α, the null hypothesis is rejected. The data supports the conclusion that the true mean cholesterol level of people who attempt suicide is less than the mean cholesterol level of people who do not attempt suicide.

11.21 Let μ_1 denote the mean frequency of alcohol use for those that rush a sorority and μ_2 denote the mean frequency of alcohol use for those that do not rush a sorority.

$H_o: \mu_1 - \mu_2 = 0 \quad H_a: \mu_1 - \mu_2 > 0$

$\alpha = 0.01$

Test statistic: $t = \dfrac{(\bar{x}_1 - \bar{x}_2) - 0}{\sqrt{\dfrac{s_1^2}{n_1} + \dfrac{s_2^2}{n_2}}}$

Assumptions: The sample size for each group is large (greater than or equal to 30) and the two samples are independently selected random samples.

$n_1 = 54$, $\bar{x}_1 = 2.72$, $s_1 = 0.86$, $n_2 = 51$, $\bar{x}_2 = 2.11$, $s_2 = 1.02$

$$t = \frac{(2.72 - 2.11) - 0}{\sqrt{\dfrac{(0.86)^2}{54} + \dfrac{(1.02)^2}{51}}} = \frac{0.61}{0.184652} = 3.30$$

$$df = \frac{\left(\dfrac{s_1^2}{n_1} + \dfrac{s_2^2}{n_2}\right)^2}{\dfrac{1}{n_1 - 1}\left(\dfrac{s_1^2}{n_1}\right)^2 + \dfrac{1}{n_2 - 1}\left(\dfrac{s_2^2}{n_2}\right)^2} = \frac{(0.0137 + 0.0204)^2}{\dfrac{(0.0137)^2}{53} + \dfrac{(0.0204)^2}{50}} = 98.002$$

So $df = 98$ (rounded down to an integer)

P-value = area under the 98 df t curve to the right of $3.30 = 1 - 0.9993 = 0.0007$

Since the P-value is less than α, the null hypothesis is rejected. The data supports the conclusion that the true mean frequency of alcohol use is larger for those that rushed a sorority than for those who did not rush a sorority.

11.23 Let μ_1 denote the mean half-life of vitamin D in plasma for people on a normal diet. Let μ_2 denote the mean half-life of vitamin D in plasma for people on a high-fiber diet. Let $\mu_1 - \mu_2$ denote the true difference in mean half-life of vitamin D in plasma for people in these two groups (normal minus high fiber).

$H_o: \mu_1 - \mu_2 = 0$ $H_a: \mu_1 - \mu_2 > 0$

$\alpha = 0.01$

Test statistic: $t = \dfrac{(\bar{x}_1 - \bar{x}_2) - 0}{\sqrt{\dfrac{s_1^2}{n_1} + \dfrac{s_2^2}{n_2}}}$

Assumptions: The population distributions are (at least approximately) normal and the two samples are independently selected random samples.

Refer to the Minitab output given in the problem statement.

From the Minitab output the P-value = 0.007. Since the P-value is less than α, H_o is rejected. There is sufficient evidence to conclude that the mean half-life of vitamin D is longer for those on a normal diet than for those on a high-fiber diet.

11.25 Let μ_1 denote the mean self-esteem score for students classified as having short duration loneliness. Let μ_2 denote the mean self-esteem score for students classified as having long duration loneliness.

$H_o: \mu_1 - \mu_2 = 0 \quad H_a: \mu_1 - \mu_2 > 0$

$\alpha = 0.01$

Test statistic: $\quad t = \dfrac{(\overline{x}_1 - \overline{x}_2) - 0}{\sqrt{\dfrac{s_1^2}{n_1} + \dfrac{s_2^2}{n_2}}}$

Assumptions: The population distributions are (at least approximately) normal and the two samples are independently selected random samples.

$n_1 = 72, \ \overline{x}_1 = 76.78 \ , s_1 = 17.8, \ n_2 = 17, \ \overline{x}_2 = 64.00, \ s_2 = 15.68$

$$t = \frac{(76.78 - 64.00) - 0}{\sqrt{\dfrac{(17.8)^2}{72} + \dfrac{(15.68)^2}{17}}} = \frac{12.78}{4.34316} = 2.9426$$

$$df = \frac{\left(\dfrac{s_1^2}{n_1} + \dfrac{s_2^2}{n_2}\right)^2}{\dfrac{1}{n_1 - 1}\left(\dfrac{s_1^2}{n_1}\right)^2 + \dfrac{1}{n_2 - 1}\left(\dfrac{s_2^2}{n_2}\right)^2} = \frac{(4.4006 + 14.4625)^2}{\dfrac{(4.4006)^2}{71} + \dfrac{(14.4625)^2}{16}} = 26.7$$

So $df = 26$ (rounded down to an integer)

P-value = area under the 26 df t curve to the right of $2.9426 \approx 0.0034$.

Since the P-value is less than α, H_o is rejected. The sample data supports the conclusion that the mean self esteem is lower for students classified as having long duration loneliness than for students classified as having short duration loneliness.

11.27 Let μ_1 denote the mean number of goals scored per game for games in which Gretzky played and μ_2 the mean number of goals scored per game for games in which he did not play.

$H_o: \mu_1 - \mu_2 = 0 \quad H_a: \mu_1 - \mu_2 > 0$

$\alpha = 0.01$

Test statistic: $t = \dfrac{(\bar{x}_1 - \bar{x}_2) - 0}{\sqrt{\dfrac{s_1^2}{n_1} + \dfrac{s_2^2}{n_2}}}$

Assumptions: The population distributions are (at least approximately) normal and the two samples are independently selected random samples.

$n_1 = 41, \bar{x}_1 = 4.73, s_1 = 1.29, n_2 = 17, \bar{x}_2 = 3.88, s_2 = 1.18$

$$t = \frac{(4.73 - 3.88) - 0}{\sqrt{\dfrac{(1.29)^2}{41} + \dfrac{(1.18)^2}{17}}} = \frac{0.85}{0.3500} = 2.4286$$

$$df = \frac{\left(\dfrac{s_1^2}{n_1} + \dfrac{s_2^2}{n_2}\right)^2}{\dfrac{1}{n_1 - 1}\left(\dfrac{s_1^2}{n_1}\right)^2 + \dfrac{1}{n_2 - 1}\left(\dfrac{s_2^2}{n_2}\right)^2} = \frac{(0.040588 + 0.081906)^2}{\dfrac{(0.040588)^2}{40} + \dfrac{(0.081906)^2}{16}} = 32.6$$

So $df = 32$ (rounded down to an integer)

P-value = area under the 32 df t curve to the right of $2.4286 \approx 0.0105$.

Since the P-value exceeds α, H_o is not rejected. At a significance level of 0.01, the sample data does not support the conclusion that the mean number of goals scored per game is larger when Gretzky played than when he didn't play.

11.29   Let μ_1 denote the mean right leg strength for males and μ_2 the mean right leg strength for females.

$n_1 = 13, \bar{x}_1 = 2127, s_1 = 513, n_2 = 14, \bar{x}_2 = 1843, s_2 = 496$

$$V_1 = \frac{s_1^2}{n_1} = \frac{(513)^2}{13} = 20243.76923 \qquad V_2 = \frac{s_2^2}{n_2} = \frac{(446)^2}{14} = 14208.28571$$

$$df = \frac{(V_1 + V_2)^2}{\dfrac{V_1^2}{n_1 - 1} + \dfrac{V_2^2}{n_2 - 1}} = \frac{(20243.76923 + 14208.28571)^2}{\dfrac{(20243.76923)^2}{12} + \dfrac{(14208.28571)^2}{13}} = 23.9$$

Use df = 23.

The 95% confidence interval for $\mu_1 - \mu_2$ based on this sample is

$(2127 - 1843) \pm 2.069\sqrt{20243.76923 + 14208.28571} \implies 284 \pm 2.069(185.613)$
$\implies 284 \pm 384.033 \implies (-100.03, 668.03).$

Based on this sample data, the difference between mean right leg strength of males and females is plausibly as large as 668.03 or as small as -100.03.

11.31 **a** Let μ_1 denote the true average oxygen consumption for courting pairs and μ_2 the true average oxygen consumption for non-courting pairs.

$H_o: \mu_1 - \mu_2 = 0$ $H_a: \mu_1 - \mu_2 > 0$

$\alpha = 0.05$

Test statistic: $t = \dfrac{(\overline{x}_1 - \overline{x}_2) - 0}{\sqrt{\dfrac{s_p^2}{n_1} + \dfrac{s_p^2}{n_2}}}$

Assumptions: The population distributions are (at least approximately) normal, the two population standard deviations are equal, and the two samples are independently selected random samples.

df for pooled t-test = $n_1 + n_2 - 2 = 11 + 15 - 2 = 24$

$s_p^2 = \dfrac{10(0.0066)^2 + 14(0.0071)^2}{24} = 0.000047555$

$t = \dfrac{(0.099 - 0.072) - 0}{\sqrt{\dfrac{0.000047555}{11} + \dfrac{0.000047555}{15}}} = \dfrac{0.027}{0.002737} = 9.86$

P-value = area under the 24 df t curve to the right of $9.86 \approx 0.0000$.

Since the P-value is less than α, H_o is rejected. There is sufficient evidence in the sample data to conclude that the true average oxygen consumption for courting pairs is larger than the true average oxygen consumption for non-courting pairs.

b Let μ_1 denote the true average oxygen consumption for courting pairs and μ_2 the true average oxygen consumption for non-courting pairs.

$H_o: \mu_1 - \mu_2 = 0$ $H_a: \mu_1 - \mu_2 > 0$

$\alpha = 0.05$

Test statistic: $t = \dfrac{(\bar{x}_1 - \bar{x}_2) - 0}{\sqrt{\dfrac{s_1^2}{n_1} + \dfrac{s_2^2}{n_2}}}$

Assumptions: The population distributions are (at least approximately) normal and the two samples are independently selected random samples.

$n_1 = 15, \bar{x}_1 = 0.099, s_1 = 0.0071, n_2 = 11, \bar{x}_2 = 0.072, s_2 = 0.0066$

$$t = \frac{(0.099 - 0.072) - 0}{\sqrt{\dfrac{(0.0071)^2}{15} + \dfrac{(0.0066)^2}{11}}} = \frac{0.027}{0.0027057} = 9.979$$

$$df = \frac{\left(\dfrac{s_1^2}{n_1} + \dfrac{s_2^2}{n_2}\right)^2}{\dfrac{1}{n_1 - 1}\left(\dfrac{s_1^2}{n_1}\right)^2 + \dfrac{1}{n_2 - 1}\left(\dfrac{s_2^2}{n_2}\right)^2} = \frac{(0.00000336 + 0.00000396)^2}{\dfrac{(0.00000336)^2}{14} + \dfrac{(0.00000396)^2}{10}} = 22.57$$

So df = 22 (rounded down to an integer)

P-value = area under the 22 df t curve to the right of 9.98 $\approx$ 0.

Since the P-value is less than α, H_o is rejected. There is sufficient evidence in the sample data to conclude that the true average oxygen consumption for courting pairs is larger than the true average oxygen consumption for non-courting pairs.

So the conclusion is the same as in part **a**.

Section 11.2

11.33   Take n pieces of pipe and cut each into two pieces, resulting in n pairs of pipe. Coat one piece in each pair with coating 1 and the other piece with coating 2. Then put both pipes from a pair into service where they are buried at the same depth, orientation, in the same soil type, etc. After the specified length of time, measure the depth of corrosion penetration for each piece of pipe. The experiment results in paired data, which "filters out" effects due to the extraneous factors.

11.35   **a**   The data are paired because the response for the number of science courses each girl in the sample intended to take is logically matched with the same girl's response for the number of science courses she thought boys should take.

b   Let μ_d denote the true average difference in the intended number of courses for girls and boys (girls – boys).

The 95% confidence interval for μ_d is

$$\bar{d} \pm (t\,\text{critical})\frac{s_d}{\sqrt{n}} \Rightarrow -0.83 \pm (1.971)\left(\frac{1.51}{\sqrt{223}}\right)$$

$$\Rightarrow -0.83 \pm 0.1988 \Rightarrow (-1.029, -0.631).$$

With 95% confidence, it is estimated that the mean difference in the number of science courses girls intend to take and what they think boys should take is −1.029 and −0.631.

11.37 Let μ_d denote the true average difference in yield between the two varieties of wheat (Sundance − Manitou).

H_o: $\mu_d = 0$ (no difference in average yield)

H_a: $\mu_d > 0$ (average yield for Sundance is larger than average yield for Manitou)

$\alpha = 0.01$

The test statistic is: $t = \dfrac{\bar{x}_d - 0}{\dfrac{s_d}{\sqrt{n}}}$ with d.f. = 8

The differences are: 815, 1084, 681, 550, 535, 786, 1162, 517, 910.
From these: $\bar{x}_d = 782.222$ and $s_d = 236.736$

$$t = \frac{782.222 - 0}{\dfrac{236.736}{\sqrt{9}}} = 9.91$$

P-value = area under the 8 df t curve to the right of $9.91 \approx 0$.

Since the P-value is less than α, the null hypothesis is rejected. The data supports the conclusion that the mean yield of Sundance exceeds the mean yield of Manitou.

11.39 Let μ_d denote the true average difference in number of seeds detected by the two methods (Direct − Stratified).

H_o: $\mu_d = 0$ (no difference in average number of seeds detected)

H_a: $\mu_d \neq 0$ (average number of seeds detected by the Direct method is not the same as the average number of seeds detected by the Stratified method)

$\alpha = 0.05$

The test statistic is: $t = \dfrac{\bar{x}_d - 0}{\dfrac{s_d}{\sqrt{n}}}$ with d.f. = 26

The differences are: 16, −4, −8, 4, −32, 0, 12, 0, 4, −8, 4, 12, 8, −28, 4, 0, 0, 4, 0, −8, −8, 0, 0, −4, −28, 4, −36.

From these: $\bar{x}_d = -3.407$ and $s_d = 13.253$

$$t = \dfrac{-3.407 - 0}{\dfrac{13.253}{\sqrt{27}}} = -1.34$$

P-value = 2(area under the 26 df t curve to the left of −1.34) ≈ 2(0.096) = 0.192. Since the P-value exceeds α, the null hypothesis is not rejected. The data do not provide sufficient evidence to conclude that the mean number of seeds detected differs for the two methods.

11.41 **a** Let μ_d denote the mean difference in blood pressure (dental setting minus medical setting).

$H_o: \mu_d = 0$ $H_a: \mu_d > 0$

$\alpha = 0.01$

The test statistic is: $t = \dfrac{\bar{x}_d - 0}{\dfrac{s_d}{\sqrt{n}}}$ with d.f. = 59

$$t = \dfrac{4.47 - 0}{\dfrac{8.77}{\sqrt{60}}} = 3.95$$

P-value = area under the 59 df t curve to the right of 3.95 ≈ 0.

Since the P-value is less than α, H_o is rejected. Thus, the data does suggest that true mean blood pressure is higher in a dental setting than in a medical setting.

b Let μ_d denote the true mean difference in pulse rate (dental minus medical).

$H_o: \mu_d = 0$ $H_a: \mu_d \neq 0$

$\alpha = 0.05$

The test statistic is: $t = \dfrac{\bar{x}_d - 0}{\dfrac{s_d}{\sqrt{n}}}$ with d.f. = 59

$$t = \frac{-1.33 - 0}{\frac{8.84}{\sqrt{60}}} = -1.165$$

P-value = 2(area under the 59 df t curve to the left of −1.165) = 2(0.124) = 0.248.

Since the P-value exceeds α, H_o is not rejected. There is not sufficient evidence to conclude that mean pulse rates differ for a dental setting and a medical setting.

11.43 Let μ_d denote the true mean difference in improvement scores (experimental minus control).

H_o: $\mu_d = 0$ H_a: $\mu_d > 0$
$\alpha = 0.10$

The test statistic is: $t = \dfrac{\overline{x}_d - 0}{\frac{s_d}{\sqrt{n}}}$ with d.f. = 6

The differences are: −0.3, −0.1, 0.7, −0.1, 1.1, −1.4, 0.2.

From these: $\overline{x}_d = 0.0143$ and $s_d = 0.7967$

$$t = \frac{0.0143 - 0}{\frac{0.7967}{\sqrt{7}}} = 0.047$$

P-value = area under the 6 df t curve to the right of 0.047 $\approx$ 0.48.

Since the P-value exceeds α, H_o is not rejected. The experimental training method does not appear to be superior to the standard training method.

Section 11.3

11.45 Let π_1 denote the proportion of students who registered by phone that were satisfied with the registration process and π_2 denote the corresponding proportion for those who registered on-line.

H_o: $\pi_1 - \pi_2 = 0$ H_a: $\pi_1 - \pi_2 < 0$

$\alpha = 0.05$

$$z = \frac{p_1 - p_2}{\sqrt{\dfrac{p_c(1-p_c)}{n_1} + \dfrac{p_c(1-p_c)}{n_2}}}$$

$$p_1 = \frac{57}{80} = 0.7125 \qquad p_2 = \frac{50}{60} = 0.8333$$

$$p_c = \frac{n_1 p_1 + n_2 p_2}{n_1 + n_2} = \frac{57 + 50}{80 + 60} = 0.7643$$

$$z = \frac{(0.7125 - 0.8333)}{\sqrt{\dfrac{0.7643(1-0.7643)}{80} + \dfrac{0.7643(1-0.7643)}{60}}} = \frac{-0.1208}{0.0725} = -1.666$$

P-value = Area under the z curve to the left of −1.666 = 0.0479.

Since the P-value is less than α, H_o is rejected. The data supports the claim that the proportion of satisfied students is higher for those who registered on-line than for those who registered over the phone.

11.47 Let π_1 denote the proportion of female Indian False Vampire bats that spend over five minutes in the air before locating food. Let π_2 denote the proportion of male Indian False Vampire bats that spend over five minutes in the air before locating food.

$$H_o: \pi_1 - \pi_2 = 0 \qquad H_a: \pi_1 - \pi_2 \neq 0$$

$$\alpha = 0.01$$

$$z = \frac{p_1 - p_2}{\sqrt{\dfrac{p_c(1-p_c)}{n_1} + \dfrac{p_c(1-p_c)}{n_2}}}$$

$$p_1 = \frac{36}{193} = 0.1865, \quad p_2 = \frac{64}{168} = 0.3810,$$

$$p_c = \frac{n_1 p_1 + n_2 p_2}{n_1 + n_2} = \frac{36 + 64}{193 + 168} = 0.277$$

$$z = \frac{(0.1865 - 0.3810)}{\sqrt{\dfrac{0.277(0.723)}{193} + \dfrac{0.277(0.723)}{168}}} = \frac{-0.1945}{0.0472} = -4.12$$

P-value = 2(area under the z curve to the left of -4.12) ≈ 2(0) = 0.

Since the P-value is less than α, H_o is rejected. There is sufficient evidence in the data to support the conclusion that the proportion of female Indian False Vampire bats who spend over five minutes in the air before locating food differs from that of male Indian False Vampire bats.

11.49 Let π_1 denote the proportion of females who are concerned about getting AIDS and let π_2 denote the proportion of males who are similarly concerned.

$$H_o: \pi_1 - \pi_2 = 0 \quad H_a: \pi_1 - \pi_2 > 0$$

We will compute P-value for this test.

$$z = \frac{p_1 - p_2}{\sqrt{\dfrac{p_c(1-p_c)}{n_1} + \dfrac{p_c(1-p_c)}{n_2}}}$$

$p_1 = 0.427, \quad p_2 = 0.275,$

$$p_c = \frac{n_1 p_1 + n_2 p_2}{n_1 + n_2} = \frac{568(0.427) + 234(0.275)}{568 + 234} = 0.3827$$

$$z = \frac{(0.427 - 0.275)}{\sqrt{\dfrac{0.3827(1-0.3827)}{568} + \dfrac{0.3827(1-0.3827)}{234}}} = \frac{0.152}{0.0378} = 4.026$$

P-value = area under the z curve to the right of $4.026 \approx 0.00003$.

Since the P-value is much smaller than any of the commonly used significance values, H_o is rejected. There is sufficient evidence in the data to support the conclusion that the proportion of females who are concerned about getting AIDS is greater than the proportion of males so concerned.

11.51 Let π_1 denote the proportion of students in the College of Computing who lose their HOPE scholarship at the end of the first year and let π_2 denote the proportion of students in the Ivan Allen College who lose their HOPE scholarship at the end of the first year.

$$H_o: \pi_1 - \pi_2 = 0 \quad H_a: \pi_1 - \pi_2 \neq 0$$

We will compute a P-value for this test.

$$z = \frac{p_1 - p_2}{\sqrt{\dfrac{p_c(1-p_c)}{n_1} + \dfrac{p_c(1-p_c)}{n_2}}}$$

$p_1 = 0.532, \quad p_2 = 0.649,$

$$P_c = \frac{n_1 p_1 + n_2 p_2}{n_1 + n_2} = \frac{137(0.532) + 111(0.649)}{137 + 111} = 0.5842$$

$$z = \frac{(0.532 - 0.649)}{\sqrt{\dfrac{0.5842(1 - 0.5842)}{137} + \dfrac{0.5842(1 - 0.5842)}{111}}} = \frac{-0.11665}{0.06294} = -1.853$$

P-value = 2 times (the area under the z curve to the left of -1.853) ≈ 0.0638.

H_o cannot be rejected at a significance level of 0.05 or smaller. There is not sufficient evidence in the data to support the conclusion that the proportion of students in the College of Computing who lose their HOPE scholarship at the end of one year is different from the proportion for the Ivan Allen College.

11.53 Let π_1 denote the proportion of priests in 1985 who agreed that celibacy should be a matter of personal choice. Let π_2 denote the proportion of priests in 1993 who agreed that celibacy should be a matter of personal choice.

$H_o: \pi_1 - \pi_2 = 0 \quad H_a: \pi_1 - \pi_2 > 0$

$\alpha = 0.05$

$$z = \frac{p_1 - p_2}{\sqrt{\dfrac{P_c(1 - P_c)}{n_1} + \dfrac{P_c(1 - P_c)}{n_2}}}$$

$n_1 = 200, \ p_1 = 0.69, \ n_2 = 200, \ p_2 = 0.38$

$$P_c = \frac{n_1 p_1 + n_2 p_2}{n_1 + n_2} = \frac{200(0.69) + 200(0.38)}{200 + 200} = 0.535$$

$$z = \frac{(0.69 - 0.38)}{\sqrt{\dfrac{0.535(0.465)}{200} + \dfrac{0.535(0.465)}{200}}} = \frac{0.31}{0.049877} = 6.21$$

P-value = area under the z curve to the right of 6.21 ≈ 0.

Since the P-value is less than α, H_o is rejected. The sample data supports the conclusion that the proportion of priests who agree that celibacy should be a matter of personal choice has declined from 1985 to 1993.

11.55 Let π_1 denote the true proportion of returning students who do not take an orientation course and π_2 denote the true proportion of returning students who do take an orientation course.

$$n_1 = 94, \ x_1 = 50, \ p_1 = \frac{50}{94} = 0.5319, \ n_2 = 94, \ x_2 = 56, \ p_2 = \frac{56}{94} = 0.5957$$

The 95% confidence interval for $\pi_1 - \pi_2$ is

$$(0.5319 - 0.5957) \pm 1.96 \sqrt{\frac{0.5319(0.4681)}{94} + \frac{0.5957(0.4043)}{94}} \Rightarrow -0.0638 \pm 1.96(0.0722)$$

$$\Rightarrow -0.0638 \pm 0.1415 \Rightarrow (-0.2053, 0.0777).$$

With 95% confidence, it is estimated that the difference between the proportion of returning who do not take an orientation course and the proportion of returning students who do take an orientation course may be as small as −0.2053 to as large as 0.0777.

11.57 Let π_1 denote the true proportion of children drinking fluoridated water who have decayed teeth, and let π_2 denote the true proportion of children drinking non-fluoridated water who have decayed teeth.

$$n_1 = 119 \ x_1 = 67 \ p_1 = \frac{67}{119} = 0.5630 \ n_2 = 143 \ x_2 = 106 \ p_2 = \frac{106}{143} = 0.7413$$

The 90% confidence interval for $\pi_1 - \pi_2$ is

$$(0.5630 - 0.7413) \pm 1.645 \sqrt{\frac{0.5630(0.4370)}{119} + \frac{0.7413(0.2587)}{143}} \Rightarrow -0.1783 \pm 1.645(0.0584)$$

$$\Rightarrow -0.1783 \pm 0.096 \Rightarrow (-0.2743, -0.0823).$$

The interval does not contain 0, so we can conclude that the two true proportions differ. Since both endpoints of the interval are negative, this indicates that $\pi_1 < \pi_2$. Thus with 90% confidence, it is estimated that the percentage of children drinking fluoridated water that have decayed teeth is less than that for children drinking non-fluoridated water by as little as about 8%, to as much as 27%.

Section 11.4

11.59 Let μ_1 denote the true average fluoride concentration for livestock grazing in the polluted region and μ_2 denote the true average fluoride concentration for livestock grazing in the unpolluted regions.

$$H_o: \mu_1 - \mu_2 = 0 \quad H_a: \mu_1 - \mu_2 > 0$$

$$\alpha = 0.05$$

The test statistic is: rank sum for polluted area (sample 1).

Sample	Ordered Data	Rank
2	14.2	1
1	16.8	2
1	17.1	3
2	17.2	4
2	18.3	5
2	18.4	6
1	18.7	7
1	19.7	8
2	20.0	9
1	20.9	10
1	21.3	11
1	23.0	12

Rank sum $= (2 + 3 + 7 + 8 + 10 + 11 + 12) = 53$

P-value: This is an upper-tail test. With $n_1 = 7$ and $n_2 = 5$, Appendix Table VI tells us that the P-value > 0.05.

Since the P-value exceeds α, H_o is not rejected. The data does not support the conclusion that there is a larger average fluoride concentration for the polluted area than for the unpolluted area.

11.61   **a**   Let μ_1 denote the true average ascent time using the lateral gait and μ_2 denote the true average ascent time using the four-beat diagonal gait.

$H_o: \mu_1 - \mu_2 = 0$ $H_a: \mu_1 - \mu_2 \neq 0$

$\alpha = 0.05$ (A value for α was not specified in the problem, so this value was chosen for illustration.)

The test statistic is: Rank sum for diagonal gait.

Gait	Ordered Data	Rank
D	0.85	1
L	0.86	2
L	1.09	3
D	1.24	4
D	1.27	5
L	1.31	6
L	1.39	7
D	1.45	8
L	1.51	9
L	1.53	10
L	1.64	11
D	1.66	12
D	1.82	13

Rank sum = 1 + 4 + 5 + 8 + 12 + 13 = 43

P-value: This is a two-tail test. With $n_1 = 7$ and $n_2 = 6$, Appendix Table VI tells us that the P-value > 0.05.

Since the P-value exceeds α, H_o is not rejected. The data does not suggest that there is a difference in mean ascent time for the diagonal and lateral gaits.

b   We can be at least 95% confident (actually 96.2% confident) that the difference in the mean ascent time using lateral gait and the mean ascent time using diagonal gait may be as small as −.43 to as large as 0.3697.

11.63   Let μ_1 denote the true mean number of binges per week for people who use Imipramine and μ_2 the true mean number of binges per week for people who use a placebo.

$H_o: \mu_1 - \mu_2 = 0$ $H_a: \mu_1 - \mu_2 < 0$

$\alpha = 0.05$

The test statistic is: Rank sum for the Imipramine group.

Group	Ordered Data	Rank
I	1	1.5
I	1	1.5
I	2	3.5
I	2	3.5
I	3	6
P	3	6
P	3	6
P	4	8.5
P	4	8.5
I	5	10
P	6	11
I	7	12
P	8	13
P	10	14
I	12	15
P	15	16

Rank sum = 1.5 + 1.5 + 3.5 + 3.5 + 6 + 10 + 12 + 15 = 53

P-value: This is an lower-tail test. With $n_1 = 8$ and $n_2 = 8$, Appendix Table VI tells us that the P-value > 0.05.

Since the P-value exceeds α, H_o is not rejected. The data does not provide enough evidence to suggest that Imipramine is effective in reducing the mean number of binges per week.

11.65 Let μ_1 denote the mean burn time for oak and μ_2 the mean burn time for pine. The approximate 95% distribution-free confidence interval for the difference of the mean burning times of oak and pine is from −0.4998 to 0.5699. The confidence interval indicates that the mean burning time of oak may be as much as 0.5699 hours longer than pine, but also that the mean burning time of oak may be as much as 0.4998 hours shorter than pine.

Supplementary Exercises

11.67 μ_1 = mean hostility score for children of divorced parents

μ_2 = mean hostility score for children of married parents

$\mu_1 - \mu_2$ = difference in mean hostility scores

$H_o: \mu_1 - \mu_2 = 0$ $\quad$ $H_a: \mu_1 - \mu_2 > 0$

$\alpha = 0.01$ (A value for α was not specified in the problem, so this value was chosen.)

Test statistic: $\quad t = \dfrac{(\bar{x}_1 - \bar{x}_2) - 0}{\sqrt{\dfrac{s_1^2}{n_1} + \dfrac{s_2^2}{n_2}}}$

Assumptions: The sample sizes for each group is large (greater than or equal to 30) and the two samples are independently selected random samples.

$n_1 = 54$, $\bar{x}_1 = 5.38$, $s_1 = 3.96$, $n_2 = 54$, $\bar{x}_2 = 1.94$, $s_2 = 3.10$

$$t = \frac{(5.38 - 1.94) - 0}{\sqrt{\dfrac{(3.96)^2}{54} + \dfrac{(3.10)^2}{54}}} = \frac{3.44}{0.684} = 5.03$$

$$df = \frac{\left(\dfrac{s_1^2}{n_1} + \dfrac{s_2^2}{n_2}\right)^2}{\dfrac{1}{n_1 - 1}\left(\dfrac{s_1^2}{n_1}\right)^2 + \dfrac{1}{n_2 - 1}\left(\dfrac{s_2^2}{n_2}\right)^2} = \frac{(0.29040 + 0.17796)^2}{\dfrac{(0.29040)^2}{53} + \dfrac{(0.17796)^2}{53}} = 100.22$$

So $df = 100$ (rounded down to an integer)

Since this is an upper-tailed test, the associated P-value is the area under the 100 df t curve to the right of the computed t value of 5.03. Since this area is approximately 0, P-value ≈ 0.

The P-value is smaller than my selected significance level of 0.01, so we reject H_0. This data does support the researcher's hypothesis that the mean hostility score is higher for children of divorced parents than for children of married parents.

11.69   **a**   Let μ_1 denote the mean extroversion score of those born under water signs and μ_2 the mean extroversion score of those born under other signs.

$H_0: \mu_1 - \mu_2 = 0$ $H_a: \mu_1 - \mu_2 < 0$

$\alpha = 0.01$

Test statistic: $t = \dfrac{(\bar{x}_1 - \bar{x}_2) - 0}{\sqrt{\dfrac{s_1^2}{n_1} + \dfrac{s_2^2}{n_2}}}$

Assumptions: The sample sizes for each group is large (greater than or equal to 30) and the two samples are independently selected random samples.

$n_1 = 59$, $\bar{x}_1 = 11.71$, $s_1 = 3.69$, $n_2 = 186$, $\bar{x}_2 = 12.53$, $s_2 = 4.14$

$$t = \frac{(11.71 - 12.53) - 0}{\sqrt{\dfrac{(3.69)^2}{59} + \dfrac{(4.14)^2}{186}}} = \frac{-0.82}{\sqrt{.2308 + .0921}} = \frac{-0.82}{0.5683} = -1.44$$

$$df = \frac{\left(\dfrac{s_1^2}{n_1} + \dfrac{s_2^2}{n_2}\right)^2}{\dfrac{1}{n_1 - 1}\left(\dfrac{s_1^2}{n_1}\right)^2 + \dfrac{1}{n_2 - 1}\left(\dfrac{s_2^2}{n_2}\right)^2} = \frac{(0.23078 + 0.09218)^2}{\dfrac{(0.23078)^2}{58} + \dfrac{(0.09218)^2}{185}} = 108.2$$

So $df = 108$ (rounded down to an integer)

P-value = area under the 108 df t curve to the left of $-1.44 = 0.07596$.

Since the P-value exceeds α, H_0 is not rejected. There is insufficient evidence to conclude that the mean extroversion score of those born under water signs is smaller than the mean extroversion score of those born under other signs.

b   Let μ_1 denote the mean extroversion score of those born under winter signs and μ_2 the mean extroversion score of those born under summer signs.

$H_0: \mu_1 - \mu_2 = 0$ $H_a: \mu_1 - \mu_2 < 0$

$\alpha = 0.05$

Test statistic: $t = \dfrac{(\bar{x}_1 - \bar{x}_2) - 0}{\sqrt{\dfrac{s_1^2}{n_1} + \dfrac{s_2^2}{n_2}}}$

Assumptions: The sample sizes for each group is large (greater than or equal to 30) and the two samples are independently selected random samples.

$n_1 = 73$, $\bar{x}_1 = 11.49$, $s_1 = 4.28$, $n_2 = 49$, $\bar{x}_2 = 13.57$, $s_2 = 3.71$

$$t = \frac{(11.49 - 13.57) - 0}{\sqrt{\dfrac{(4.28)^2}{73} + \dfrac{(3.71)^2}{49}}} = \frac{-2.08}{\sqrt{0.2509 + 0.2809}} = \frac{-2.08}{0.7292} = -2.85$$

$$df = \frac{\left(\dfrac{s_1^2}{n_1} + \dfrac{s_2^2}{n_2}\right)^2}{\dfrac{1}{n_1 - 1}\left(\dfrac{s_1^2}{n_1}\right)^2 + \dfrac{1}{n_2 - 1}\left(\dfrac{s_2^2}{n_2}\right)^2} = \frac{(0.25094 + 0.28090)^2}{\dfrac{(0.25094)^2}{72} + \dfrac{(0.28090)^2}{48}} = 112.3$$

So $df = 112$ (rounded down to an integer)

P-value = area under the 112 df t curve to the left of $-2.85 = 0.0026$.

Since the P-value is less than α, H_0 is rejected. Thus, the data does suggest that those born under winter signs have a lower mean extroversion score than those born under summer signs.

c   Let μ_1 denote the mean neuroticism score of those born under water signs and μ_2 denote the mean neuroticism score of those born under other signs.

$H_0: \mu_1 - \mu_2 = 0$ $H_a: \mu_1 - \mu_2 \neq 0$

$\alpha = 0.05$

Test statistic: $t = \dfrac{(\bar{x}_1 - \bar{x}_2) - 0}{\sqrt{\dfrac{s_1^2}{n_1} + \dfrac{s_2^2}{n_2}}}$

Assumptions: The sample sizes for each group is large (greater than or equal to 30) and the two samples are independently selected random samples.

$n_1 = 59$, $\bar{x}_1 = 12.32$, $s_1 = 4.15$, $n_2 = 186$, $\bar{x}_2 = 12.23$, $s_2 = 4.11$

$$t = \frac{(12.32 - 12.23) - 0}{\sqrt{\dfrac{(4.15)^2}{59} + \dfrac{(4.11)^2}{186}}} = \frac{0.09}{\sqrt{0.2919 + 0.0908}} = \frac{0.09}{0.6186} = 0.145$$

$$df = \frac{\left(\dfrac{s_1^2}{n_1} + \dfrac{s_2^2}{n_2}\right)^2}{\dfrac{1}{n_1 - 1}\left(\dfrac{s_1^2}{n_1}\right)^2 + \dfrac{1}{n_2 - 1}\left(\dfrac{s_2^2}{n_2}\right)^2} = \frac{(0.29191 + 0.09081)^2}{\dfrac{(0.29191)^2}{58} + \dfrac{(0.09081)^2}{185}} = 96.8$$

So $df = 96$ (rounded down to an integer)

P-value $= 2$(area under the 96 df t curve to the right of 0.145) $= 2(1 - 0.5577)$
$ = 0.8846.$

Since the P-value exceeds α, the null hypothesis is not rejected. It appears that the mean neuroticism score for those born under water signs does not differ from that of those born under other signs.

d   Let μ_1 denote the mean neuroticism score for people born under winter signs and μ_2 denote the mean neuroticism score for people born under summer signs.

H_o: $\mu_1 - \mu_2 = 0$ H_a: $\mu_1 - \mu_2 \neq 0$

$\alpha = 0.01$

Test statistic: $t = \dfrac{(\bar{x}_1 - \bar{x}_2) - 0}{\sqrt{\dfrac{s_1^2}{n_1} + \dfrac{s_2^2}{n_2}}}$

Assumptions: The sample sizes for each group is large (greater than or equal to 30) and the two samples are independently selected random samples.

$n_1 = 73$, $\bar{x}_1 = 11.96$, $s_1 = 4.22$, $n_2 = 49$, $\bar{x}_2 = 13.27$, $s_2 = 4.04$

$$t = \frac{(11.96 - 13.27) - 0}{\sqrt{\dfrac{(4.22)^2}{73} + \dfrac{(4.04)^2}{49}}} = \frac{-1.31}{\sqrt{.2440 + .3331}} = \frac{-1.31}{\sqrt{.577}} = \frac{-1.31}{.7596} = -1.72$$

$$df = \frac{\left(\dfrac{s_1^2}{n_1} + \dfrac{s_2^2}{n_2}\right)^2}{\dfrac{1}{n_1 - 1}\left(\dfrac{s_1^2}{n_1}\right)^2 + \dfrac{1}{n_2 - 1}\left(\dfrac{s_2^2}{n_2}\right)^2} = \frac{(0.24395 + 0.33309)^2}{\dfrac{(0.24395)^2}{72} + \dfrac{(0.33309)^2}{48}} = 106.1$$

So $df = 106$ (rounded down to an integer)

P-value = 2(area under the 106 df t curve to the left of -1.72) = 2(0.04377) = 0.08754.

Since the P-value of 0.08754 is greater than the α value of 0.01, H_o cannot be rejected. The data do not suggest that the mean neuroticism score for those born under winter signs differ from the mean neuroticism score for those born under summer signs.

11.71 Let μ_1 denote the true mean final exam scores for students who take quizzes at the beginning of the class and μ_2 the true mean final exam scores for students who take quizzes at the end of class.

$H_o: \mu_1 - \mu_2 = 0 \quad H_a: \mu_1 - \mu_2 \neq 0$

No level of significance was specified, so $\alpha = 0.01$ will be used for illustration.

Test statistic: $\quad t = \dfrac{(\bar{x}_1 - \bar{x}_2) - 0}{\sqrt{\dfrac{s_1^2}{n_1} + \dfrac{s_2^2}{n_2}}}$

Assumptions: The sample sizes for each group is large (greater than or equal to 30) and the two samples are independently selected random samples.

$n_1 = 40$, $\bar{x}_1 = 143.7$, $s_1 = 21.2$, $n_2 = 40$, $\bar{x}_2 = 131.7$, $s_2 = 20.9$

$$t = \frac{(143.7 - 131.7) - 0}{\sqrt{\dfrac{(21.2)^2}{40} + \dfrac{(20.9)^2}{40}}} = \frac{12}{\sqrt{11.236 + 10.92}} = \frac{12}{4.71} = 2.55$$

$$df = \frac{\left(\dfrac{s_1^2}{n_1} + \dfrac{s_2^2}{n_2}\right)^2}{\dfrac{1}{n_1 - 1}\left(\dfrac{s_1^2}{n_1}\right)^2 + \dfrac{1}{n_2 - 1}\left(\dfrac{s_2^2}{n_2}\right)^2} = \frac{(11.2360 + 10.9203)^2}{\dfrac{(11.2360)^2}{39} + \dfrac{(10.9203)^2}{39}} = 77.98$$

So $df = 77$ (rounded down to an integer)

P-value = 2(area under the 77 df t curve to the right of 2.55) = 2(1 − 0.9936) = 0.0128.

Since the P-value exceeds α, H_o is not rejected. There is insufficient evidence to conclude that the true mean exam scores for the two groups differ.

11.73 Let μ_1 denote the true mean self-esteem score for students who are members of Christian groups and μ_2 the true mean self-esteem score for students who are not members of Christian groups.

$H_o: \mu_1 - \mu_2 = 0$ $H_a: \mu_1 - \mu_2 \neq 0$

$\alpha = 0.01$

Test statistic: $t = \dfrac{(\overline{x}_1 - \overline{x}_2) - 0}{\sqrt{\dfrac{s_1^2}{n_1} + \dfrac{s_2^2}{n_2}}}$

Assumptions: The sample sizes for each group is large (greater than or equal to 30) and the two samples are independently selected random samples.

$n_1 = 169$, $\overline{x}_1 = 25.08$, $s_1 = 10$, $n_2 = 124$, $\overline{x}_2 = 24.55$, $s_2 = 8$

$t = \dfrac{(25.08 - 24.55) - 0}{\sqrt{\dfrac{(10)^2}{169} + \dfrac{(8)^2}{124}}} = \dfrac{0.53}{1.052542} = 0.5035$

$df = \dfrac{\left(\dfrac{s_1^2}{n_1} + \dfrac{s_2^2}{n_2}\right)^2}{\dfrac{1}{n_1 - 1}\left(\dfrac{s_1^2}{n_1}\right)^2 + \dfrac{1}{n_2 - 1}\left(\dfrac{s_2^2}{n_2}\right)^2} = \dfrac{(0.59172 + 0.51613)^2}{\dfrac{(0.59172)^2}{168} + \dfrac{(0.51613)^2}{123}} = 288.8$

So $df = 288$ (rounded down to an integer)

P-value = 2(area under the 288 df t curve to the right of 0.5035) = 2(0.3075) = 0.6150.

Since the P-value exceeds α, the null hypothesis is not rejected. The sample data does not support the conclusion that the mean self-esteem score for students who are members of Christian groups differs from that of students who are not members of Christian groups.

11.75   a   Let μ_1 denote the mean number of lightning flashes for single-peak storms and μ_2 the mean number of lightning flashes for multiple-peak storms.

$H_o: \mu_1 - \mu_2 = 0$ $H_a: \mu_1 - \mu_2 \neq 0$

$\alpha = 0.05$

Test statistic: $t = \dfrac{(\overline{x}_1 - \overline{x}_2) - 0}{\sqrt{\dfrac{s_1^2}{n_1} + \dfrac{s_2^2}{n_2}}}$

Assumptions: Each population is (at least approximately) normally distributed and the two samples are independently selected random samples.

From the data given, the following statistics were calculated.

$$n_1 = 7, \bar{x}_1 = 66.143, s_1 = 31.746, \ n_2 = 4, \bar{x}_2 = 274.5, s_2 = 105.38$$

$$t = \frac{(66.143 - 274.5)}{\sqrt{\dfrac{(31.746)^2}{7} + \dfrac{(105.38)^2}{4}}} = \frac{-208.357}{54.0375} = -3.856$$

$$df = \frac{\left(\dfrac{s_1^2}{n_1} + \dfrac{s_2^2}{n_2}\right)^2}{\dfrac{1}{n_1-1}\left(\dfrac{s_1^2}{n_1}\right)^2 + \dfrac{1}{n_2-1}\left(\dfrac{s_2^2}{n_2}\right)^2} = \frac{(143.9729 + 2776.0825)^2}{\dfrac{(143.9729)^2}{6} + \dfrac{(2776.0825)^2}{3}} = 3.3$$

So $df = 3$ (rounded down to an integer)

P-value = 2(area under the 3 df t curve to the left of −3.856) ≈ 2(0.0154) = 0.0308.

Since the P-value is less than α, H$_o$ is rejected. The sample data supports the conclusion that there is a difference between the mean number of lightening flashes for single-peak and multiple-peak storms.

b   Assumption: The distributions of the number of lightning flashes of single-peak storms and multiple-peak storms are normal.

11.77 Let μ_1 denote the mean ratio for young men and μ_2 the mean ratio for elderly men.

$H_o: \mu_1 - \mu_2 = 0$ $H_a: \mu_1 - \mu_2 > 0$

$\alpha = 0.05$

Test statistic: $t = \dfrac{(\bar{x}_1 - \bar{x}_2) - 0}{\sqrt{\dfrac{s_1^2}{n_1} + \dfrac{s_2^2}{n_2}}}$

Assumptions: The two populations of ratios are normally distributed and the two samples are independently selected random samples.

From the data given, the following statistics were calculated.

$$n_1 = 13, \bar{x}_1 = 7.47, \frac{s_1}{\sqrt{n_1}} = 0.22, s_1 = 0.793221, \ n_2 = 12, \bar{x}_2 = 6.71, \frac{s_2}{\sqrt{n_2}} = 0.28, s_2 = 0.969948$$

$$t = \frac{(7.47 - 6.71)}{\sqrt{\frac{(0.793221)^2}{13} + \frac{(0.969948)^2}{12}}} = \frac{0.76}{0.3561} = 2.1343$$

$$df = \frac{\left(\frac{s_1^2}{n_1} + \frac{s_2^2}{n_2}\right)^2}{\frac{1}{n_1 - 1}\left(\frac{s_1^2}{n_1}\right)^2 + \frac{1}{n_2 - 1}\left(\frac{s_2^2}{n_2}\right)^2} = \frac{(0.0484 + 0.0784)^2}{\frac{(0.0484)^2}{12} + \frac{(0.0784)^2}{11}} = 21.3$$

So $df = 21$ (rounded down to an integer)

P-value = area under the 21 df t curve to the right of $2.1343 \approx 0.0223$.

Since the P-value is less than α, H_o is rejected. The sample data supports the conclusion that the mean ratio for young men exceeds that for elderly men.

11.79 Let μ_1 and μ_2 denote the mean nitrogen concentration in soil 0.6m and 6m, respectively, from the roadside.

$H_o: \mu_1 - \mu_2 = 0$ $H_a: \mu_1 - \mu_2 > 0$

$\alpha = 0.01$

Test statistic: $t = \frac{(\bar{x}_1 - \bar{x}_2) - 0}{\sqrt{\frac{s_1^2}{n_1} + \frac{s_2^2}{n_2}}}$

Assumptions: The two populations are normally distributed and the two samples are independently selected random samples.

$n_1 = 20, \bar{x}_1 = 1.7, s_1 = 0.4, n_2 = 20, \bar{x}_2 = 1.35, s_2 = 0.3$

$$t = \frac{(1.7 - 1.35)}{\sqrt{\frac{(0.4)^2}{20} + \frac{(0.3)^2}{20}}} = \frac{0.35}{0.111803} = 3.13$$

$$df = \frac{\left(\frac{s_1^2}{n_1} + \frac{s_2^2}{n_2}\right)^2}{\frac{1}{n_1 - 1}\left(\frac{s_1^2}{n_1}\right)^2 + \frac{1}{n_2 - 1}\left(\frac{s_2^2}{n_2}\right)^2} = \frac{(0.0080 + 0.0045)^2}{\frac{(0.0080)^2}{19} + \frac{(0.0045)^2}{19}} = 35.2$$

So $df = 35$ (rounded down to an integer)

P-value = area under the 35 df t curve to the right of $3.13 \approx 0.0018$.

Since the P-value is less than α, the null hypothesis is rejected. The data does provide sufficient evidence to conclude that the mean nitrogen concentration in soil 0.6m from the roadside is higher than in soil 6m from the roadside.

11.81　**a**　Let μ_1 denote the true mean birth weight for premature infants with brain damage and μ_2 the true mean birth weight for premature infants without brain damage.

$H_o: \mu_1 - \mu_2 = 0 \quad H_a: \mu_1 - \mu_2 \neq 0$

$\alpha = 0.05$

Test statistic: $\quad t = \dfrac{(\overline{x}_1 - \overline{x}_2) - 0}{\sqrt{\dfrac{s_1^2}{n_1} + \dfrac{s_2^2}{n_2}}}$

Assumptions: The two populations are normally distributed and the two samples are independently selected random samples.

$n_1 = 10,\ \overline{x}_1 = 1541,\ s_1 = 448,\ n_2 = 54,\ \overline{x}_2 = 1204,\ s_2 = 652$

$$t = \frac{(1541 - 1204)}{\sqrt{\dfrac{(448)^2}{10} + \dfrac{(652)^2}{54}}} = \frac{337}{167.16069} = 2.016$$

$$df = \frac{\left(\dfrac{s_1^2}{n_1} + \dfrac{s_2^2}{n_2}\right)^2}{\dfrac{1}{n_1 - 1}\left(\dfrac{s_1^2}{n_1}\right)^2 + \dfrac{1}{n_2 - 1}\left(\dfrac{s_2^2}{n_2}\right)^2} = \frac{(20070.4 + 7872.296)^2}{\dfrac{(20070.4)^2}{9} + \dfrac{(7872.296)^2}{53}} = 17.0007$$

So $df = 17$ (rounded down to an integer)

P-value = 2(area under the 17 df t curve to the right of 2.016) $\approx 2(0.03) = 0.06$.

Since the P-value is less than α, the null hypothesis is rejected. There is not sufficient evidence to conclude that there is a difference in the mean birth weight of infants born premature with brain damage and those born premature with no brain damage.

b　Let μ_1 denote the true mean birth weight for full-term infants with brain damage and μ_2 the true mean birth weight for full-term infants without brain damage.

$H_o: \mu_1 - \mu_2 = 0 \quad H_a: \mu_1 - \mu_2 \neq 0$

$\alpha = 0.05$

Test statistic: $\quad t = \dfrac{(\bar{x}_1 - \bar{x}_2) - 0}{\sqrt{\dfrac{s_1^2}{n_1} + \dfrac{s_2^2}{n_2}}}$

Assumptions: The two populations are normally distributed and the two samples are independently selected random samples.

$n_1 = 12, \ \bar{x}_1 = 2998 \ , s_1 = 707, \ n_2 = 13, \ \bar{x}_2 = 2704, \ s_2 = 627$

$$t = \frac{(2998 - 2704)}{\sqrt{\dfrac{(707)^2}{12} + \dfrac{(627)^2}{13}}} = \frac{294}{268.132012} = 1.096$$

$$df = \frac{\left(\dfrac{s_1^2}{n_1} + \dfrac{s_2^2}{n_2}\right)^2}{\dfrac{1}{n_1 - 1}\left(\dfrac{s_1^2}{n_1}\right)^2 + \dfrac{1}{n_2 - 1}\left(\dfrac{s_2^2}{n_2}\right)^2} = \frac{(41654.083 + 30240.692)^2}{\dfrac{(41654.083)^2}{11} + \dfrac{(30240.692)^2}{12}} = 22.09$$

So $df = 22$ (rounded down to an integer)

P-value = 2(area under the 22 df t curve to the right of 1.096) $\approx$ 2(0.142) = 0.284.

Since the P-value exceeds α, the null hypothesis is not rejected. There is not sufficient evidence to conclude that there is a difference in the mean birth weight of infants carried full term with brain damage and those carried full term without brain damage.

c   Let μ_1 denote the mean birth weight of infants without brain damage carried full term and μ_2 the mean birth weight of infants without brain damage born premature.

$n_1 = 13, \ \bar{x}_1 = 2704 \ , s_1 = 627, \ n_2 = 54, \ \bar{x}_2 = 1204, \ s_2 = 656$

$$V_1 = \frac{s_1^2}{n_1} = \frac{(627)^2}{13} = 30240.692308 \qquad V_2 = \frac{s_2^2}{n_2} = \frac{(656)^2}{54} = 7969.185185$$

$$df = \frac{(V_1 + V_2)^2}{\dfrac{V_1^2}{n_1 - 1} + \dfrac{V_2^2}{n_2 - 1}} = \frac{(30240.692308 + 7969.185185)^2}{\dfrac{(30240.692308)^2}{12} + \dfrac{(7969.185185)^2}{53}} = 18.86$$

Use df = 18

The 90% confidence interval for $\mu_1 - \mu_2$ based on this sample is

$$(2704 - 1204) \pm 1.73\sqrt{30240.692308 + 7969.185185}$$
$$\Rightarrow \quad 1500 \pm 1.73(195.47347) \quad \Rightarrow \quad 1500 \pm 338.17 \quad \Rightarrow \quad (1161.83, \ 1838.17).$$

d   Let μ_1 denote the mean birth weight of infants with brain damage carried full term and μ_2 the mean birth weight of infants with brain damage born premature.

$$n_1 = 12, \ \bar{x}_1 = 2998, \ s_1 = 707, \ n_2 = 10, \ \bar{x}_2 = 1541, \ s_2 = 448$$

$$V_1 = \frac{s_1^2}{n_1} = \frac{(707)^2}{12} = 41654.083333 \qquad V_2 = \frac{s_2^2}{n_2} = \frac{(448)^2}{10} = 20070.4$$

$$df = \frac{(V_1 + V_2)^2}{\dfrac{V_1^2}{n_1 - 1} + \dfrac{V_2^2}{n_2 - 1}} = \frac{(41654.083333 + 20070.4)^2}{\dfrac{(41654.083333)^2}{11} + \dfrac{(20070.4)^2}{9}} = 18.8$$

Use df = 18

The 90% confidence interval for $\mu_1 - \mu_2$ based on this sample is

$$(2998 - 1541) \pm 1.734\sqrt{41654.083333 + 20070.4}$$
$$\Rightarrow \quad 1457 \pm 1.734(248.444125) \Rightarrow 1457 \pm 430.802 \Rightarrow (1026.20, 1887.80).$$

11.83   The large sample z test for testing a difference of two proportions requires that the samples be independent. In this survey, the same people answered the initial question and the revised question and so the two samples are not independent. There is no procedure from this chapter that can be used to answer the question posed.

11.85 **a**   Let μ_d denote the true mean difference in pH level (surface pH minus subsoil pH).

The differences are: $-0.23, -0.16, -0.21, 0.26, -0.18, 0.17, 0.25, -0.20$.

From these: $\bar{x}_d = -0.0375$ and $s_d = 0.2213$

The 90% confidence interval for μ_d is

$$\bar{d} \pm (t\,critical)\frac{s_d}{\sqrt{n}} \Rightarrow -.0375 \pm (1.8946)\frac{.2213}{\sqrt{8}}$$

$$\Rightarrow -0.0375 \pm 0.1482 \Rightarrow (-.1857, .1107).$$

With 90% confidence, it is estimated that the mean difference in pH is between $-.1857$ and 0.1107.

b   The assumption made about the underlying pH distributions is that each distribution is normal, so the distribution of differences is normal.

11.87   Let μ_1 denote the true mean age at death for female SIDS victims and μ_2 the true mean age at death for male SIDS victims.

From the data,

$$n_1 = 5, \ \bar{x}_1 = 103.6, \ s_1^2 = 2154.3, \ n_2 = 7, \ \bar{x}_2 = 89.7, \ s_2^2 = 1638.238$$

$$V_1 = \frac{s_1^2}{n_1} = \frac{(2154.3)}{5} = 430.86 \quad V_2 = \frac{s_2^2}{n_2} = \frac{(1638.238)}{7} = 234.034$$

$$df = \frac{(V_1 + V_2)^2}{\dfrac{V_1^2}{n_1 - 1} + \dfrac{V_2^2}{n_2 - 1}} = \frac{(430.86 + 234.034)^2}{\dfrac{(430.86)^2}{4} + \dfrac{(234.034)^2}{6}} = 7.9599$$

Use df = 7.

The confidence interval is

$$(103.6 - 89.7) \pm (2.365)\sqrt{430.86 + 234.034} \Rightarrow 13.9 \pm (2.365)(25.7855)$$

$$\Rightarrow 13.9 \pm 60.98 \Rightarrow (-47.08, 74.88).$$

With 95% confidence, it is estimated that $\mu_1 - \mu_2$ is between −47.08 and 74.88. Since this interval contains zero, the analysis supports the conclusion that there may be no difference between the true mean ages at death of male and female SIDS victims.

11.89   Let μ_d denote the true mean difference in LH release (before minus after).

$$H_o: \mu_d = 0 \quad H_a: \mu_d > 0$$

$$\alpha = 0.01$$

The test statistic is:

$$t = \frac{\bar{x}_d - 0}{\dfrac{s_d}{\sqrt{n}}} \quad \text{with d.f.} = 6$$

The differences are: 150, −25, 275, 525, 350, 225, 150.

From these: $\bar{x}_d = 235.71$ and $s_d = 173.72$

$$t = \frac{235.71 - 0}{\frac{173.72}{\sqrt{7}}} = 3.59$$

P-value = area under the 6 df t curve to the right of 3.59 = 0.006.

Since the P-value is less than α, H_o is rejected. The data supports the conclusion that there is a significant reduction in mean LH release following treatment with ACTH.

Chapter 12
The Analysis of Categorical Data
and Goodness-of-Fit Tests

Section 12.1

12.1 **a** $0.020 < \text{P-value} < 0.025$

b $0.040 < \text{P-value} < 0.045$

c $0.035 < \text{P-value} < 0.040$

d $\text{P-value} < 0.001$

e $\text{P-value} > 0.100$

12.3 **a** $df = 3$ and $X^2 = 19.0$. From Appendix Table IX, P-value < 0.001. Since the P-value is less than α, H_o is rejected.

b If $n = 40$, then it is not advisable to use the chi-square test since one of the expected cell frequencies (cell corresponding to nut type 4) would be less than 5.

12.5 Let π_i denote the true proportion of birds choosing color i first ($i = 1, 2, 3, 4$). Here 1=Blue, 2=Green, 3=Yellow, and 4=Red.

H_o: $\pi_1 = \pi_2 = \pi_3 = \pi_4 = 0.25$

H_a: at least one of the true proportions differ from 0.25.

$\alpha = 0.01$

Test statistic: $X^2 = \sum \dfrac{(\text{observed count} - \text{expected count})^2}{\text{expected count}}$

$n = 16 + 8 + 6 + 3 = 33$.

Expected count for each cell = 33(0.25) = 8.25.

$$X^2 = \frac{(16-8.25)^2}{8.25} + \frac{(8-8.25)^2}{8.25} + \frac{(6-8.25)^2}{8.25} + \frac{(3-8.25)^2}{8.25}$$
$$= 7.28030 + 0.00758 + 0.61364 + 3.34091 = 11.242$$

df = 3. From Appendix Table IX, 0.010 < P-value < 0.015. Since the P-value is greater than α, H_o is not rejected. The data do not provide sufficient evidence indicating a color preference.

12.7 **a** Let π_i denote the true proportion of policy holders in Astrological sign group i. (i = 1 for Aquarius, i = 2 for Aries, ..., i = 12 for Virgo).

H_o: $\pi_1 = \pi_2 = \pi_3 = \pi_4 = \pi_5 = \pi_6 = \pi_7 = \pi_8 = \pi_9 = \pi_{10} = \pi_{11} = \pi_{12} = 1/12$

H_a: at least one of the true proportions differ from 1/12

$\alpha = 0.05$ (A significance level is not specified in this problem. We use $\alpha = 0.05$ for illustration.)

Test statistic: $X^2 = \sum \dfrac{(\text{observed count} - \text{expected count})^2}{\text{expected count}}$

n = 35,666 + 37,926 + 38,126 + 54,906 + 37,179 + 37,354 + 37,910 + 36,677

+ 34,175 + 35,352 + 37,179 + 37,718 = 460,168

Expected count for each cell = 460,168(1/12) = 38,347.3.

$$X^2 = \frac{(35,666-38,347.3)^2}{38,347.3} + \frac{(37,926-38,347.3)^2}{38,347.3} + \frac{(38,126-38,347.3)^2}{38,347.3}$$
$$+ \frac{(54,906-38,347.3)^2}{38,347.3} + \frac{(37,179-38,347.3)^2}{38,347.3} + \frac{(37,354-38,347.3)^2}{38,347.3} +$$
$$\frac{(37,910-38,347.3)^2}{38,347.3} + \frac{(36,677-38,347.3)^2}{38,347.3} + \frac{(34,175-38,347.3)^2}{38,347.3}$$
$$+ \frac{(35,352-38,347.3)^2}{38,347.3} + \frac{(37,179-38,347.3)^2}{38,347.3} + \frac{(37,718-38,347.3)^2}{38,347.3}$$

=187.48 + 4.63 + 1.28 + 7150.16 + 35.60 + 25.73 + 4.99 + 72.76 + 453.97 + 233.97

+ 35.60 + 10.33 = 8,216.48

df =11. From Appendix Table IX, P-value < 0.001. Since the P-value is less than than α, H_o is rejected.

The data provide very strong evidence to conclude that the proportions of policy holders are not all equal for the twelve astrological signs.

b   Sign of Capricorn covers birthdates between December 22 and January 20 which is the summer season in Australia. One possible explanation for the higher than expected proportion of policy holders for this sign might be that more teenagers start driving during the summer months than any other months and hence more policies are issued during this period.

c   Let π_i denote the true proportion of policy holders in Astrological sign group i who make claims. (i = 1 for Aquarius, i = 2 for Aries, ..., i = 12 for Virgo).

H_o: $\pi_1 = 35666/460168 = 0.077506$, $\pi_2 = 37926/460168 = 0.082418$,
$\pi_3 = 38126/460168 = 0.082852$, $\pi_4 = 54906/460168 = 0.119317$,
$\pi_5 = 37179/460168 = 0.080794$, $\pi_6 = 37354/460168 = 0.081175$,
$\pi_7 = 37910/460168 = 0.082383$, $\pi_8 = 36677/460168 = 0.079703$,
$\pi_9 = 34175/460168 = 0.074266$, $\pi_{10} = 35352/460168 = 0.076824$,
$\pi_{11} = 37179/460168 = 0.080794$, $\pi_{12} = 37718/460168 = 0.081966$

H_a: at least one of the true proportions differs from the hypothesized value.

$\alpha = 0.01$ (A significance level is not specified in this problem. We use $\alpha = 0.01$ for illustration.)

Test statistic: $X^2 = \sum \dfrac{(\text{observed count} - \text{expected count})^2}{\text{expected count}}$

n = 1000

The required calculations for obtaining the expected cell frequencies are summarized in the table below. The number of policy holders of the company for the different astrological signs is given in the second column of the table. The corresponding proportions are given in the third column. The number of policy holders in the sample making claims for the different astrological signs is given in column 4. The corresponding expected number of claims is given in the last column. The expected number is calculated as follows:

$$\left(\begin{array}{c}\text{Expected number of claims}\\ \text{for this astrological sign}\end{array}\right) = 1000 \times \frac{\text{Number of policy holders with this sign}}{\text{Total number of policy holders}}$$

Sign	Number of policy holders	Proportion of policy holders	Number of claims in the sample	Expected number of claims in the sample
Aquarius	35666	0.077506	85	77.506
Aries	37926	0.082418	83	82.418
Cancer	38126	0.082852	82	82.852
Capricorn	54906	0.119317	88	119.317
Gemini	37179	0.080794	83	80.794
Leo	37354	0.081175	83	81.175
Libra	37910	0.082383	83	82.383
Pisces	36677	0.079703	82	79.703
Sagittarius	34175	0.074266	81	74.266
Scorpio	35352	0.076824	85	76.824
Taurus	37179	0.080794	84	80.794
Virgo	37718	0.081966	81	81.966

$$X^2 = \frac{(85-77.506)^2}{77.506} + \frac{(83-82.418)^2}{82.418} + \frac{(82-82.852)^2}{82.852} + \frac{(88-119.317)^2}{119.317}$$

$$+ \frac{(83-80.794)^2}{80.794} + \frac{(83-81.175)^2}{81.175} + \frac{(83-82.383)^2}{82.383} + \frac{(82-79.703)^2}{79.703}$$

$$+ \frac{(81-74.266)^2}{74.266} + \frac{(85-76.824)^2}{76.824} + \frac{(84-80.794)^2}{80.794} + \frac{(81-81.966)^2}{81.966}$$

$$= 0.72449 + 0.00411 + 0.00877 + 8.21987 + 0.06021 + 0.04104 + 0.00462 + 0.06617 + 0.61053$$

$$+ 0.87011 + 0.12718 + 0.01138 = 10.7485$$

df $=11$. From Appendix Table IX, P-value > 0.45. At a significance level of $\alpha = 0.01$, H_o cannot be rejected. It cannot be rejected even at a significance level of $\alpha = 0.05$. The data do not provide evidence to conclude that the proportions of claims are consistent with the proportions of policy holders for the various astrological signs. However, it is worth noting that the proportion of claims by policy holders in the sample belonging to Capricorn is much smaller than what would be expected based on the overall proportion of policy holders with this sign!

12.9 Let $\pi_1, \pi_2, \pi_3, \pi_4, \pi_5$ denote the true proportions of the various offense categories.

H_o: $\quad \pi_1 = 0.307$, $\pi_2 = 0.386$, $\pi_3 = 0.093$, $\pi_4 = 0.206$, $\pi_5 = 0.008$

H_a: $\quad H_o$ is not true.

$\alpha = 0.05$

Test statistic: $X^2 = \sum \dfrac{(\text{observed count} - \text{expected count})^2}{\text{expected count}}$

Computations:

n = 999

Offense	Violent crime	Crimes against property	Drug related	Public order offense	Other
Frequency	225	300	230	228	16
Expected	306.693	385.614	92.907	205.794	7.992

$$X^2 = \frac{(225 - 306.693)^2}{306.693} + \frac{(300 - 385.614)^2}{385.614} + \frac{(230 - 92.907)^2}{92.907}$$

$$+ \frac{(228 - 205.794)^2}{205.794} + \frac{(16 - 7.992)^2}{7.992}$$

$$= 21.760 + 19.008 + 202.294 + 2.396 + 8.024 = 253.482$$

df = 4. From Appendix Table IX, P-value < 0.001. Since the P-value is less than α, the null hypothesis is rejected. The data does provide sufficient evidence to conclude that the true 1989 proportions falling into the various offense categories are not the same as in 1983.

12.11 Let π_i denote the proportion of all returned questionnaires accompanied by cover letter i (i = 1, 2, 3).

H_o: $\pi_1 = \frac{1}{3}, \pi_2 = \frac{1}{3}, \pi_3 = \frac{1}{3}$

H_a: H_o is not true.

$\alpha = 0.05$

Test statistic: $X^2 = \sum \frac{(\text{observed count} - \text{expected count})^2}{\text{expected count}}$

Computations: n = 131 $\frac{n}{3}$ = 43.67

$$X^2 = \frac{(48 - 43.67)^2}{43.67} + \frac{(44 - 43.67)^2}{43.67} + \frac{(39 - 43.67)^2}{43.67} = 0.429 + 0.0025 + 0.499 = 0.931$$

df = 2. From Appendix Table IX, P-value > 0.100. Since the P-value exceeds α, the null hypothesis is not rejected. The data does not suggest that the proportions of returned questionnaires differ for the three cover letters.

12.13   Let π_i denote the proportion of phenotype i (i = 1, 2, 3, 4).

H_o: $\pi_1 = \dfrac{9}{16}, \pi_2 = \dfrac{3}{16}, \pi_3 = \dfrac{3}{16}, \pi_4 = \dfrac{1}{16}$

H_a: H_o is not true

$\alpha = 0.01$

Test statistic: $X^2 = \sum \dfrac{(\text{observed count} - \text{expected count})^2}{\text{expected count}}$

Computations:

	\multicolumn{4}{c}{Phenotype}	
	1	2	3	4	Total
Frequency	926	288	293	104	16.11
Expected	906.19	302.06	302.06	100.69	

$X^2 = \dfrac{(926-906.19)^2}{906.19} + \dfrac{(288-302.06)^2}{302.06} + \dfrac{(293-302.06)^2}{302.06} + \dfrac{(104-100.69)^2}{100.69}$

$= 0.433 + 0.655 + 0.278 + 0.109 = 1.47$

df = 3. From Appendix Table IX, P-value > 0.10. Since the P-value exceeds α, the null hypothesis is not rejected. The data appears to be consistent with Mendel's laws.

Section 12.2

12.15   **a**   The d.f. will be (6 - 1)(3 - 1) = 10.

b   The d.f. will be (7 - 1)(3 - 1) = 12.

c   The d.f. will be (6 - 1)(4 - 1) = 15.

12.17 H_o: the response category proportions are the same for the two cover designs.

H_a: the proportions are not the same for all response categories for the two cover designs.

$\alpha = 0.05$

Test statistic: $X^2 = \sum \dfrac{(\text{observed count} - \text{expected count})^2}{\text{expected count}}$

	1 - 7	8 - 14	15 – 31	32 - 60	Not returned	
Graphic	70 (75.8)	76 (63.5)	51 (50.2)	19 (25.1)	198 (199.4)	414
Plain	84 (78.2)	53 (65.5)	51 (51.8)	32 (25.9)	207 (205.6)	427
	154	129	102	51	405	841

Since all expected cell counts are at least 5, the X^2 statistics can be used.

$df = (2 - 1)(5 - 1) = 4$

$$X^2 = \frac{(70-75.8)^2}{75.8} + \frac{(76-63.5)^2}{63.5} + \frac{(51-50.2)^2}{50.2} + \frac{(19-25.1)^2}{25.1} + \frac{(198-199.4)^2}{199.4}$$

$$+ \frac{(84-78.2)^2}{78.2} + \frac{(53-65.5)^2}{65.5} + \frac{(51-51.8)^2}{51.8} + \frac{(32-25.9)^2}{25.9} + \frac{(207-205.6)^2}{205.6}$$

$$= 0.4438 + 2.4606 + 0.0127 + 1.4825 + 0.0098 + 0.4302 + 2.3855 + 0.0124 + 1.4367 + 0.0095 = 8.6837$$

From Appendix Table IX, 0.065 < P-value < 0.070. Since the P-value exceeds α, the null hypothesis is not rejected. The data does not support the theory that the proportions falling in the various response categories differ for the two cover designs. There is no evidence that cover design affects speed of response.

12.19 **a** H_o: Gender and workaholism type are independent

H_a: Gender and workaholism type are not independent

$\alpha = 0.05$ (no significance level is given in the problem so we use 0.05 for illustration.)

Test statistic: $X^2 = \sum \frac{(\text{observed count} - \text{expected count})^2}{\text{expected count}}$

Observed and expected frequencies are given in the table below (expected frequencies in parentheses)

		Female	Male	
Workaholism types	Work enthusiasts	20 (27.40)	41 (33.60)	61
	Workaholics	32 (30.99)	37 (38.01)	69
	Enthusiastic workaholics	34 (35.93)	46 (44.07)	80
	Unengaged workers	43 (42.67)	52 (52.33)	95
	Relaxed workers	24 (22.91)	27 (28.09)	51
	Disenchanted workers	37 (30.09)	30 (36.91)	67
		190	233	423

$$X^2 = \frac{(20-27.40)^2}{27.40} + \frac{(32-30.99)^2}{30.99} + \frac{(34-35.93)^2}{35.93} + \frac{(43-42.67)^2}{42.67}$$

$$+ \frac{(24-22.91)^2}{22.91} + \frac{(37-30.09)^2}{30.09} + \frac{(41-33.60)^2}{33.60} + \frac{(37-38.01)^2}{38.01}$$

$$+ \frac{(46-44.07)^2}{44.07} + \frac{(52-52.33)^2}{52.33} + \frac{(27-28.09)^2}{28.09} + \frac{(30-36.91)^2}{36.91}$$

$$= 6.852$$

$df = (6-1)(2-1) = 5$. From Appendix Table IX, P-value > 0.10. Hence the null hypothesis is not rejected. The data are consistent with the hypothesis of no association between gender and workaholism type.

b   Another interpretation of the lack of association between gender and workaholism type is that, for each workaholism category, the true proportion of women who belong to this category is equal to the true proportion of men who belong to this category.

12.21 H_o: Seat location and motion sickness are independent.

H_a: Seat location and motion sickness are not independent.

$\alpha = 0.05$

Test statistic: $X^2 = \sum \frac{(\text{observed count} - \text{expected count})^2}{\text{expected count}}$

Observed and expected frequencies are given in the table below (expected frequencies in parentheses)

		Nausea	No nausea	
Seat location	Front	58 (118.85)	870 (809.15)	928
	Middle	166 (170.21)	1163 (1158.79)	1329
	Rear	193 (127.94)	806 (871.06)	999
		417	2839	3256

$$X^2 = \frac{(58-118.85)^2}{118.85} + \frac{(166-170.21)^2}{170.21} + \frac{(193-127.94)^2}{127.94} + \frac{(870-809.15)^2}{809.15}$$

$$+ \frac{(1163-1158.79)^2}{1158.79} + \frac{(806-871.06)^2}{871.06} = 73.789$$

$df = (3-1)(2-1) = 2$. From Appendix Table IX, P-value < 0.001. Hence the null hypothesis is rejected. The data provide strong evidence to conclude that seat location and nausea are dependent.

12.23 H_o: There is no dependence between the approach used and whether or not a donation is obtained.

H_a: There is a dependence between the approach used and whether or not a donation is obtained.

$\alpha = 0.05$ (No significance level is given in the problem. We use 0.05 for illustration.)

Test statistic: $X^2 = \sum \dfrac{(\text{observed count } - \text{ expected count})^2}{\text{expected count}}$

Observed and expected frequencies are given in the table below (expected frequencies in parentheses)

	Contribution made	No contribution made	
Picture of a smiling child	18 (18.33)	12 (11.67)	30
Picture of an unsmiling child	14 (18.33)	16 (11.67)	30
Verbal message	16 (15.89)	10 (10.11)	26
Identification of charity only	18 (13.44)	4 (8.56)	22
	66	42	108

$$X^2 = \frac{(18-18.33)^2}{18.33} + \frac{(14-18.33)^2}{18.33} + \frac{(16-15.89)^2}{15.89} + \frac{(18-13.44)^2}{13.44} + \frac{(12-11.67)^2}{11.67}$$

$$+ \frac{(16-11.67)^2}{11.67} + \frac{(10-10.11)^2}{10.11} + \frac{(4-8.56)^2}{8.56} = 6.621$$

df = (4 − 1)(2 − 1) = 3. From Appendix Table IX, P-value > 0.05. Hence the null hypothesis is not rejected. The data do not provide sufficient evidence to conclude that there is a dependence between the approach used to obtain donations and whether or not a donation is successfully obtained.

12.25 H_o: Position and Role are independent.

H_a: Position and Role are not independent.

$\alpha = 0.01$

Test statistic: $X^2 = \sum \dfrac{(\text{observed count} - \text{expected count})^2}{\text{expected count}}$

Observed and expected frequencies are given in the table below (expected frequencies in parentheses)

	Initiate chase	Participate in chase	
Center position	28 (39.04)	48 (36.96)	76
Wing position	66 (54.96)	41 (52.04)	107
	94	89	183

$$X^2 = \frac{(28-39.04)^2}{39.04} + \frac{(66-54.96)^2}{54.96} + \frac{(48-36.96)^2}{36.96} + \frac{(41-52.04)^2}{52.04} = 10.97$$

df = (2 − 1)(2 − 1) = 1. From Appendix Table IX, P-value < 0.001. Hence the null hypothesis is
    is
rejected. The data provide strong evidence to conclude that there is an association between position and role.

For the chi-square analysis to be valid, the observations on the 183 lionesses in the sample are assumed to be independent.

12.27 H_o: There is no dependence between response and region of residence.

H_a: There is a dependence between response and region of residence.

$\alpha = 0.01$

Test statistic: $X^2 = \sum \dfrac{(\text{observed count} - \text{expected count})^2}{\text{expected count}}$

Response

		Agree	Disagree	
	Northeast	130 (150.35)	59 (38.65)	189
Region	West	146 (149.55)	42 (38.45)	188
	Midwest	211 (209.22)	52 (53.78)	263
	South	291 (268.88)	47 (69.12)	338
		778	200	978

$$X^2 = \frac{(130-150.35)^2}{150.35} + \frac{(59-38.65)^2}{38.65} + \frac{(146-149.55)^2}{149.55} + \frac{(42-38.45)^2}{38.45}$$

$$+ \frac{(211-209.22)^2}{209.22} + \frac{(52-53.78)^2}{53.78} + \frac{(291-268.88)^2}{268.88} + \frac{(47-69.12)^2}{69.12}$$

$$= 2.754 + 10.714 + 0.084 + 0.329 + 0.015 + 0.059 + 1.820 + 7.079 = 22.855$$

df = (4 − 1)(2 − 1) = 3. From Appendix Table IX, 0.001 > P-value. Since the P-value is less than α, the null hypothesis is rejected. The data supports the conclusion that there is a dependence between response and region of residence.

12.29 H_o: The proportion of correct sex identifications is the same for each nose view.

H_a: The proportion of correct sex identifications is not the same for each nose view.

$\alpha = 0.05$

Test statistic: $X^2 = \sum \dfrac{(\text{observed count} - \text{expected count})^2}{\text{expected count}}$

		Front	Profile	Three quarter	
Sex ID	Correct	23 (26)	26 (26)	29 (26)	78
	Not Correct	17 (14)	14 (14)	11 (14)	42
		40	40	40	120

$$X^2 = \frac{(23-26)^2}{26} + \frac{(26-26)^2}{26} + \frac{(29-26)^2}{26} + \frac{(17-14)^2}{14} + \frac{(14-14)^2}{14} + \frac{(11-14)^2}{14}$$

$$= 0.346 + 0.000 + 0.346 + 0.643 + 0.000 + 0.643 = 1.978$$

df $= (2 - 1)(3 - 1) = 2$. From Appendix Table IX, P-value > 0.10. Since the P-value exceeds α, the null hypothesis is not rejected. The data does not support the hypothesis that the proportions of correct sex identifications differ for the three different nose views.

12.31 H_0: Student marijuana use is independent of parental drug and alcohol use.

H_a: Student marijuana use and parental drug and alcohol use are dependent.

$\alpha = 0.01$

Test statistic: $X^2 = \sum \frac{(\text{observed count} - \text{expected count})^2}{\text{expected count}}$

df $= 4$ and from the SAS output, the value of the chi-square statistic is read as 22.373. From Appendix Table IX, P-value ≈ 0.000. Since the P-value is less than α, H_0 is rejected. There does appear to be a dependence between student use of marijuana and parental use of alcohol and drugs.

12.33 H_0: Job satisfaction and teaching level are independent.

H_a: Job satisfaction and teaching level are dependent.

$\alpha = 0.05$

Test statistic: $X^2 = \sum \frac{(\text{observed count} - \text{expected count})^2}{\text{expected count}}$

Computations:

		Job satisfaction		
		Satisfied	Unsatisfied	
	College	74 (63.763)	43 (53.237)	117
Teaching Level	High School	224 (215.270)	171 (179.730)	395
	Elementary	126 (144.967)	140 (121.033)	266
	Total	424	354	778

$$X^2 = \frac{(74-63.763)^2}{63.763} + \frac{(43-53.237)^2}{53.237} + \frac{(224-215.270)^2}{215.270}$$

$$+ \frac{(171-179.730)^2}{179.730} + \frac{(126-144.967)^2}{144.967} + \frac{(140-121.023)^2}{121.023}$$

$$= 1.644 + 1.968 + 0.354 + 0.424 + 2.482 + 2.972 = 9.844$$

df $= (3-1)(2-1) = 2$. From Appendix Table IX, $0.010 >$ P-value > 0.005. Since the P-value is less than α, H_o is rejected. The data supports the conclusion that there is a dependence between job satisfaction and teaching level.

Supplementary Exercises

12.35 H_o: Sex and seat belt usage are independent.

H_a: Sex and seat belt usage are dependent.

$\alpha = 0.05$

Test statistic: $X^2 = \sum \dfrac{(\text{observed count} - \text{expected count})^2}{\text{expected count}}$

Computations:

	Don't use	Use	Total
Male	192 (215.69)	272 (248.31)	464
Female	284 (260.31)	276 (299.69)	560
Total	476	548	1024

$$X^2 = \frac{(192 - 215.69)^2}{215.69} + \frac{(272 - 248.31)^2}{248.31} + \frac{(284 - 260.31)^2}{260.31} + \frac{(276 - 299.69)^2}{299.69}$$

$$= 2.60 + 2.26 + 2.16 + 1.87 = 8.89$$

df = $(2 - 1)(2 - 1) = 1$. From Appendix Table IX, 0.005 > P-value > 0.001. Since the P-value is less than α, the null hypothesis is rejected. The data suggests that the variables sex and seat belt usage are dependent.

12.37 H$_o$: Sex and relative importance assigned to work and home are independent.

H$_a$: Sex and relative importance assigned to work and home are dependent.

$\alpha = 0.05$

Test statistic: $X^2 = \sum \dfrac{(\text{observed count} - \text{expected count})^2}{\text{expected count}}$

Computations:

Relative Importance

	work > home	work = home	work < home	Total
Female	68 (79.3)	26 (25.0)	94 (83.7)	188
Male	75 (63.7)	19 (20.0)	57 (67.3)	151
Total	143	45	151	339

$X^2 = 1.61 + 0.04 + 1.26 + 2.01 + 0.05 + 1.56 = 6.54$

df = $(2 - 1)(3 - 1) = 2$. From Appendix Table IX, 0.040 > P-value > 0.035. Since the P-value is less than α, the null hypothesis is rejected. The data suggests that sex and relative importance assigned to work and home are dependent.

12.39 H$_o$: Age and "Rate believed attainable" are independent.

H$_a$: Age and "Rate believed attainable" are not dependent.

$\alpha = 0.01$

Test statistic: $X^2 = \sum \dfrac{(\text{observed count} - \text{expected count})^2}{\text{expected count}}$

Computations:

Rates believed attainable

		0 - 5	6 - 10	11 - 15	Over 15	Total
	Under 45	15 (28.4)	51 (72.1)	51 (28.2)	29 (17.3)	146
	45 - 54	31 (54.8)	133 (139.3)	70 (54.5)	48 (33.4)	282
Age	55 - 64	59 (49.2)	139 (124.9)	35 (48.9)	20 (29.9)	253
	65 over	84 (56.6)	157 (143.7)	32 (56.3)	18 (34.4)	291
	Total	189	480	188	115	972

$$X^2 = \frac{(15-28.4)^2}{28.4} + \frac{(51-72.1)^2}{72.1} + \frac{(51-28.2)^2}{28.2} + \frac{(29-17.3)^2}{17.3}$$

$$+ \frac{(31-54.8)^2}{54.8} + \frac{(133-139.3)^2}{139.3} + \frac{(70-54.5)^2}{54.5} + \frac{(48-33.4)^2}{33.4}$$

$$+ \frac{(59-49.2)^2}{49.2} + \frac{(139-124.9)^2}{124.9} + \frac{(35-48.9)^2}{48.9} + \frac{(20-29.9)^2}{29.9}$$

$$+ \frac{(84-56.6)^2}{56.6} + \frac{(157-143.7)^2}{143.7} + \frac{(32-56.3)^2}{56.3} + \frac{(18-34.4)^2}{34.4}$$

$$= 6.31 + 6.17 + 18.35 + 7.96 + 10.36 + 0.28 + 4.38 + 6.42$$

$$+ 1.95 + 1.58 + 3.97 + 3.3 + 13.28 + 1.23 + 10.48 + 7.84$$

$$= 103.87$$

df = (4 − 1)(4 − 1) = 9. From Appendix Table IX, 0.001 > P-value. Since the P-value is less than α, the null hypothesis is rejected. The data very strongly suggests that the variables, age and "Rate believed attainable" are dependent.

12.41 H_o: The true proportions of individuals in each of the cocaine use categories do not differ for the three treatments.

H_a: The true proportions of individuals in each of the cocaine use categories differ for the three treatments.

$\alpha = 0.05$

Test statistic: $X^2 = \sum \dfrac{(\text{observed count} - \text{expected count})^2}{\text{expected count}}$

Computations:

Treatment

Usage	A	B	C	Total
None	149 (118.9)	75 (84.8)	8 (28.3)	232
1 – 2	26 (34.8)	27 (24.9)	15 (8.3)	68
3 – 6	6 (19.0)	20 (13.5)	11 (4.5)	37
7 or more	4 (12.3)	10 (8.8)	10 (2.9)	24
Total	185	132	44	361

$$X^2 = \frac{(149-118.9)^2}{118.9} + \frac{(75-84.8)^2}{84.8} + \frac{(8-28.3)^2}{28.3} + \frac{(26-34.8)^2}{34.8}$$

$$+\frac{(27-24.9)^2}{24.9} + \frac{(15-8.3)^2}{8.3} + \frac{(6-19)^2}{19} + \frac{(20-13.5)^2}{13.5}$$

$$+\frac{(11-4.5)^2}{4.5} + \frac{(4-12.3)^2}{12.3} + \frac{(10-8.8)^2}{8.8} + \frac{(10-2.9)^2}{2.9}$$

$= 7.62 + 1.14 + 14.54 + 2.25 + 0.18 + 5.44 + 8.86 + 3.1 + 9.34 + 5.6 + 0.17 + 17.11 = 75.35$

df $= (4 - 1)(3 - 1) = 6$. From Appendix Table IX, $0.001 > $ P-value. Since the P-value is less than α, the null hypothesis is rejected.

Note: There are two cells with expected counts of 5 or less. If you combine the usage category "3 – 6" with "7 or more," the following analysis results.

Treatment

Usage	A	B	C	Total
none	149 (118.9)	75 (84.8)	8 (28.3)	232
1 - 2	26 (34.8)	27 (24.9)	15 (8.3)	68
3 or more	10 (31.3)	30 (22.3)	21 (7.4)	61
Total	185	132	44	361

$$X^2 = \frac{(149-118.9)^2}{118.9} + \frac{(75-84.8)^2}{84.8} + \frac{(8-28.3)^2}{28.3}$$

$$+ \frac{(26-34.8)^2}{34.8} + \frac{(27-24.9)^2}{24.9} + \frac{(15-8.3)^2}{8.3}$$

$$+ \frac{(10-31.3)^2}{31.3} + \frac{(30-22.3)^2}{22.3} + \frac{(21-7.4)^2}{7.4}$$

$$= 7.62 + 1.14 + 14.54 + 2.25 + 0.18 + 5.44 + 14.46 + 2.65 + 24.75 = 73.03$$

df = (3 - 1)(3 - 1) = 4. From Appendix Table IX, 0.001 > P-value. Since the P-value is less than α, the null hypothesis is rejected.

In either analysis, the null hypothesis is rejected. The data suggests the true proportions of individuals in each of the cocaine use categories differ for the three treatments.

Chapter 13
Simple Linear Regression and Correlation:
Inferential Methods

Section 13.1

13.1 **a** $y = -5.0 + 0.017x$

   b When $x = 1000$, $y = -5 + 0.017(1000) = 12$
   When $x = 2000$, $y = -5 + 0.017(2000) = 29$

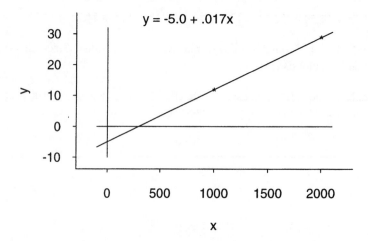

   c When $x = 2100$, $y = -5 + (0.017)(2100) = 30.7$

   d 0.017

   e $0.017(100) = 1.7$

   f It is stated that the community where the given regression model is valid has no small houses. Therefore, there is no assurance that the model is adequate for predicting usage based on house size for small houses. Consequently, it is not advisable to use the model to predict usage for a 500 sq.ft house.

13.3   **a**   The mean value of serum manganese when Mn intake is 4.0 is $-2 + 1.4(4) = 3.6$.
The mean value of serum manganese when Mn intake is 4.5 is $-2 + 1.4(4.5) = 4.3$.

     b $\dfrac{5-3.6}{1.2} = 1.17$

P(serum Mn over 5) = $P(1.17 < z) = 1 - 0.8790 = 0.121$.

     c   The mean value of serum manganese when Mn intake is 5 is $-2 + 1.4(5) = 5$.

$\dfrac{5-5}{1.2} = 0,$ $\dfrac{3.8-5}{1.2} = -1$

P(serum Mn over 5) = $P(0 < z) = 0.5$
P(serum Mn below 3.8) = $P(z < -1) = 0.1587$

13.5   **a**   The expected change in price associated with one extra square foot of space is 47. The expected change in price associated with 100 extra square feet of space is $47(100) = 4700$.

     b   When x = 1800, the mean value of y is $23000 + 47(1800) = 107600$.

$\dfrac{110000 - 107600}{5000} = 0.48$ $\dfrac{100000 - 107600}{5000} = -1.52$

P(y > 110000) = $P(0.48 < z) = 1 - 0.6844 = 0.3156$

P(y < 100000) = $P(z < -1.52) = 0.0643$

Approximately 31.56% of homes with 1800 square feet would be priced over 110,000 dollars and about 6.43% would be priced under 100,000 dollars.

13.7   **a** $r^2 = 1 - \dfrac{\text{SSResid}}{\text{SSTo}} = 1 - \dfrac{27.890}{73.937} = 1 - 0.3772 = 0.6228$

     b $s_e^2 = \dfrac{\text{SSResid}}{n-2} = 1 - \dfrac{27.890}{13-2} = \dfrac{27.890}{11} = 2.5355$

$s_e = \sqrt{2.5355} = 1.5923$

The magnitude of a typical deviation of residence half-time (y) from the population regression line is estimated to be about 1.59 hours.

     c $b = 3.4307$

     d $\hat{y} = 0.0119 + 3.4307(1) = 3.4426$

13.9   **a** $r^2 = 1 - \dfrac{2620.57}{22398.05} = 0.883$

b $s_e = \sqrt{\dfrac{2620.57}{14}} = \sqrt{187.184} = 13.682$ with 14 d f.

13.11   **a**

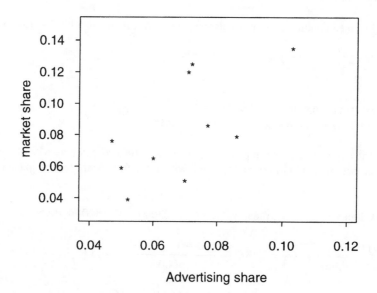

Advertising share

There seems to be a general tendency for y to increase at a constant rate as x increases. However, there is also quite a bit of variability in the y values. It is questionable whether a simple linear regression model using x is adequate for predicting. It may be advisable to look for additional predictor variables.

b   Summary values are: n = 10, $\sum x = 0.688$, $\sum x^2 = 0.050072$, $\sum y = 0.835$, $\sum y^2 = 0.079491$, $\sum xy = 0.060861$.

$$b = \frac{0.060861 - \left[\dfrac{(0.688)(0.835)}{10} \right]}{0.050072 - \left[\dfrac{(0.688)^2}{10} \right]} = \frac{0.003413}{0.0027376} = 1.246712$$

$a = 0.0835 - (1.246712)(0.0688) = -0.002274$

The equation of the estimated regression line is $\hat{y} = -0.002274 + 1.246712x$.

The predicted market share, when advertising share is 0.09, would be $-0.002274 + 1.246712(0.09) = 0.10993$.

c
$$SSTo = 0.079491 - \left[\frac{(0.835)^2}{10}\right] = 0.0097685$$

$SSResid = 0.079491 - (-0.002274)(0.835) - (1.246712)(0.060861) = 0.0055135$

$$r^2 = 1 - \frac{0.0055135}{0.0097685} = 1 - 0.564 = 0.436$$

This means that 43.6% of the total variability in market share (y) can be explained by the simple linear regression model relating market share and advertising share (x).

d
$$S_e = \sqrt{\frac{0.0055135}{8}} = \sqrt{0.000689} = 0.0263 \text{ with 8 d } f.$$

Section 13.2

13.13 **a** $\sum(x - \overline{x})^2 = 250 \quad \sigma_b = \frac{4}{\sqrt{250}} = 0.253$

b $\sum(x - \overline{x})^2 = 500 \quad \sigma_b = \frac{4}{\sqrt{500}} = 0.179$

No, the resulting value of σ_b is not half of what it was in **a**. However, the resulting value of σ_b^2 is half of what it was in **a**.

c It would require 4 observations at each x value to yield a value of σ_b which is half the value calculated in **a**. In this case $\sum(x - \overline{x})^2 = 1000$, so $\sigma_b = \frac{4}{\sqrt{1000}} = 0.1265$

13.15 **a** $S_e = \sqrt{\frac{1235.47}{13}} = \sqrt{95.036} = 9.7486$

$$S_b = \frac{9.7486}{\sqrt{4024.2}} = \frac{9.7486}{63.4366} = 0.1537$$

b The 95% confidence interval for β is $2.5 \pm (2.16)(0.1537) \Rightarrow 2.5 \pm 0.33$ $\Rightarrow (2.17, 2.83)$.

c The interval is relatively narrow. However, whether β has been precisely estimated or not depends on the particular application we have in mind.

13.17 **a** $S_{xy} = 44194 - \left[\frac{(50)(16705)}{20}\right] = 2431.5$

$$S_{xx} = 150 - \left[\frac{(50)^2}{20} \right] = 25$$

$$b = \frac{2431.5}{25} = 97.26, \qquad a = 835.25 - (97.26)(2.5) = 592.1$$

b $\hat{y} = 592.1 + 97.26(2) = 786.62$. The corresponding residual is
$(y - \hat{y}) = 757 - 786.62 = -29.62$.

c SSResid $= 14194231 - 592.1(16705) - 97.26(44194) = 4892.06$

$$s_e = \sqrt{\frac{4892.06}{18}} = \sqrt{271.781} = 16.4858$$

$$s_b = \frac{16.4858}{\sqrt{25}} = 3.2972$$

The 99% confidence interval for β, the true average change in oxygen usage associated with a one-minute increase in exercise time is

$$97.26 \pm (2.88)(3.2972) \Rightarrow 97.26 \pm 9.50 \Rightarrow (87.76, 106.76).$$

13.19 **a** $H_0: \beta = 0 \qquad H_a: \beta \neq 0$

$\alpha = 0.05$ (for illustration)

$$t = \frac{b}{s_b} \qquad \text{with df} = 42$$

$$t = \frac{15}{5.3} = 2.8302$$

P-value $= 2$(area under the 42 df t curve to the right of 2.83) $\approx 2(0.0036) = 0.0072$.

Since the P-value is less than α, H_0 is rejected. The data supports the conclusion that the simple linear regression model specifies a useful relationship between x and y. (It is advisable to examine a scatter plot of y versus x to confirm the appropriateness of a straight line model for these data).

b $b \pm (\text{t critical}) \, s_b \Rightarrow 15 \pm (2.02)(5.3) \Rightarrow 15 \pm 10.706 \Rightarrow (4.294, 25.706)$

Based on this interval, we estimate the change in mean average SAT score associated with an increase of $1000 in expenditures per child is between 4.294 and 25.706.

13.21 Summary values are: n = 10, $\sum x = 6,970$; $\sum x^2 = 5,693,950$; $\sum y = 10,148$;
$\sum y^2 = 12,446,748$; $\sum xy = 8,406,060$.

We first calculate various quantities needed to answer the different parts of this problem.

$$b = \frac{8406060 - \left[\dfrac{(6970)(10148)}{10}\right]}{5693950 - \left[\dfrac{(6970)^2}{10}\right]} = \frac{1332904}{835860} = 1.5946498217404828560$$

$$a = \frac{[1014.8 - (1.5946498217404828560)(697)]}{10} = -96.6709257531116550619$$

SSResid = 12446748 − (−96.6709257531116550619)(10148)
$\qquad\qquad$ − (1.5946498217404828560)(8406060) = 23042.474

NOTE: Using the formula $SSResid = \sum y^2 - a\sum y - b\sum xy$ can lead to severe roundoff errors unless many significant digits are carried along for the intermediate calculations. The calculations for this problem are particularly prone to roundoff errors because of the large numbers involved. This is the reason we have given many significant digits for the slope and the intercept estimates. You may want to try doing these calculations with fewer significant digits. You will notice a substantial loss in accuracy in the final answer. The formula $SSResid = \sum (y - \hat{y})^2$ provides a more numerically stable alternative. We give the calculations based on this alternative formula in the table below.

We used $b = 1.59465$ and $a = \dfrac{[1014.8 - (1.59465)(697)]}{10} = -96.6711$ to calculate $\hat{y} = a + bx$.

y	$\hat{y} = a + bx$	$y - \hat{y}$	$(y - \hat{y})^2$
303	301.99	1.009	1
491	477.4	13.597	184.9
659	660.79	-1.788	3.2
683	740.52	-57.52	3308.6
922	876.07	45.935	2110
1044	1083.37	-39.37	1550
1421	1306.62	114.379	13082.6
1329	1370.41	-41.407	1714.5
1481	1513.93	-32.925	1084.1

The sum of the numbers in the last column gives $SSResid = 23042.5$ which is accurate to the first decimal place.

$$s_e^2 = \frac{23042.5}{8} = 2880.31$$

$$s_b^2 = \frac{2880.31}{835860} = 0.00344592, \quad s_b = 0.0587020$$

a   The prediction equation is CHI = -96.6711 + 1.59465 Control .

Using this equation we can predict the mean response time for those suffering a closed-head injury using the mean response time on the same task for individuals with no head injury.

b   Let β denote the expected increase in mean response time for those suffering a closed-head injury associated with a one unit increase in mean response time for the same task for individuals with no head injury.

$H_o: \beta = 0 \quad\quad H_a: \beta \neq 0$

$\alpha = 0.05$

$$t = \frac{b}{s_b} \quad \text{with df} = 8$$

$$t = \frac{1.59465}{0.0587020} = 27.1652$$

P-value = 2(area under the 8 df t curve to the right of 27.1652) $\approx$ 2(0) = 0.

Since the P-value is less than α, H_o is rejected. The simple linear regression model does provide useful information for predicting mean response times for individuals with CHI and mean response times for the same task for individuals with no head injury. (A scatter plot of y versus x confirms that a straight line model is a reasonable model for this problem).

c   The equation CHI = 1.48 Control says that the mean response time for individuals with CHI is *proportional* to the mean response time for the same task for individuals with no head injury, and the proportionality constant is 1.48. This implies that the mean response time for individuals with CHI is *1.48 times* the mean response time for the same task for individuals with no head injury.

13.23   **a**   Let β denote the expected change in sales revenue associated with a one unit increase in advertising expenditure.

$H_o: \beta = 0 \quad\quad H_a: \beta \neq 0$

$\alpha = 0.05$

$$t = \frac{b}{s_b} \quad \text{with df} = 13$$

$$t = \frac{52.57}{8.05} = 6.53$$

P-value = 2(area under the 13 df t curve to the right of 6.53) ≈ 2(0) = 0.

Since the P-value is less than α, H_o is rejected. The simple linear regression model does provide useful information for predicting sales revenue from advertising expenditures.

b H_o: β = 40 H_a: β > 40

α = 0.01

Test statistic: $t = \frac{b - 40}{s_b}$ with df = 13

$$t = \frac{(52.57 - 40)}{8.05} = 1.56$$

P-value = area under the 13 df t curve to the right of 1.56 ≈ 0.071.

Since the P-value exceeds α, the null hypothesis is not rejected. The data are consistent with the hypothesis that the change in sales revenue associated with a one unit increase in advertising expenditure does not exceed 40 thousand dollars.

13.25   Let β denote the average change in milk pH associated with a one unit increase in temperature.

H_o: β = 0 H_a: β < 0

α = 0.01

The test statistic is: $t = \frac{b}{s_b}$ with d.f. = 14.

Computations: n = 16, $\Sigma x = 678$, $\Sigma y = 104.54$,

$$S_{xy} = 4376.36 - \frac{(678)(104.54)}{16} = -53.5225$$

$$S_{xx} = 36056 - \frac{(678)^2}{16} = 7325.75$$

$$b = \frac{-53.5225}{7325.75} = -0.0073$$

a = 6.53375 − (−0.0073)(42.375) = 6.8431

SSResid = 683.447 − 6.8431(104.54) − (−0.00730608)(4376.36) = 0.0177354

$$s_e = \sqrt{\frac{.0177354}{14}} = \sqrt{.001267} = .0356$$

$$s_b = \frac{.0356}{\sqrt{7325.75}} = .000416$$

$$t = \frac{-0.00730608}{0.000416} = -17.5627$$

P-value = area under the 14 df t curve to the left of $-17.5627 \approx 0$.

Since the P-value is less than α, H_0 is rejected. There is sufficient evidence in the sample to conclude that there is a negative (inverse) linear relationship between temperature and pH. A scatter plot of y versus x confirms this.

Section 13.3

13.27

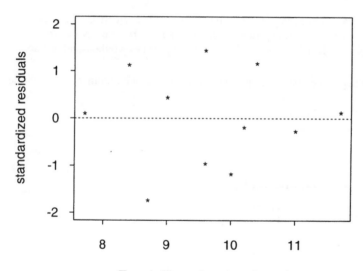

The standardized residual plot does not exhibit any unusual features.

13.29 **a** The assumptions required in order that the simple linear regression model be appropriate are:

    (i)   The distribution of the random deviation e at any particular x value has mean value 0.

    (ii)   The standard deviation of e is the same for any particular value of x.

    (iii) The distribution of e at any particular x value is normal.

(iv) The mean value of vigor is a linear function of stem density.

(v) The random deviations $e_1, e_2, \cdots, e_n$ associated with different observations are independent of one another.

b

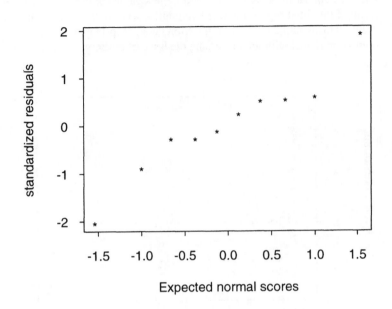

The normal probability plot appears to follow a straight line pattern (approximately). Hence the assumption that the random deviation distribution is normal is plausible.

c

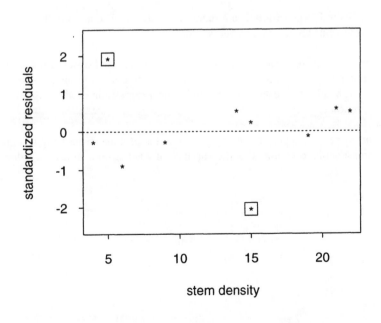

There are two residuals that are relatively large. The corresponding points are enclosed in boxes on the graph above.

d   The negative residuals appear to be associated with small x values, and the positive residuals appear to be associated with large x values. Such a pattern is apparently the result of the fitted regression line being influenced by the two potential outlying points. This would cause one to question the appropriateness of using a simple linear regression model without addressing the issue of outliers.

13.31   **a**

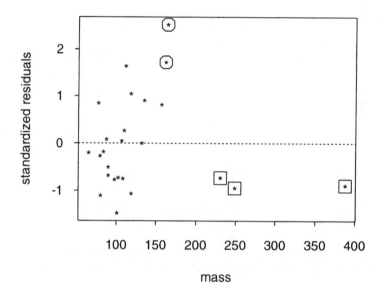

The several large residuals are marked by circles. The potentially influential observations are marked by rectangles.

b   The residuals associated with the potentially influential observations are all negative. Without these three, there appears to be a positive trend to the standardized residual plot. The plot suggests that the simple linear regression model might not be appropriate.

c   There does not appear to be any pattern in the plot that would suggest that it is unreasonable to assume that the variance of y is the same at each x value.

13.33

Year	X	Y	Y-Pred	Residual
1963	188.5	2.26	1.750	0.51000
1964	191.3	2.60	2.478	0.12200
1965	193.8	2.78	3.128	−0.34800
1966	195.9	3.24	3.674	−0.43400
1967	197.9	3.80	4.194	−0.39400
1968	199.9	4.47	4.714	−0.24400
1969	201.9	4.99	5.234	−0.24400
1970	203.2	5.57	5.572	−0.00200
1971	206.3	6.00	6.378	−0.37800
1972	208.2	5.89	6.872	−0.98200
1973	209.9	8.64	7.314	1.32600

U.S. popluation (millions)

The residuals are positive in 1963 and 1964, then they are negative from 1965 through 1972, followed by a positive residual in 1973. The residuals exhibit a pattern in the plot and thus the plot casts doubt on the appropriateness of the simple linear regression model.

Section 13.4

13.35 If the request is for a confidence interval for β the wording would likely be "estimate the change in the average y value associated with a one unit increase in the x variable." If the request is for a confidence interval for $\alpha + \beta x^*$ the wording would likely be "estimate the average y value when the value of the x variable is x^*."

13.37 **a** $S_{xx} = 2939 - \dfrac{(131)^2}{12} = 2939 - 1430.0833 = 1508.9167$

$$s_b = \frac{2.33}{\sqrt{1508.9167}} = \frac{2.33}{38.8448} = 0.059982$$

$b \pm (t\ critical)s_b \Rightarrow -0.345043 \pm (2.23)(0.059982)$
$\Rightarrow -0.345043 \pm 0.133761 \Rightarrow (-0.4788,\ -0.2113)$

b $a + b(1) = 20.125053 - 0.345043(1) = 19.78$

$$s_{a+b(1)} = 2.33\sqrt{\frac{1}{12} + \frac{(1-10.9167)^2}{1508.9167}} = 2.33(0.385365) = 0.8979$$

The 95% confidence interval is $19.78 \pm 2.23(0.8979)$

$\Rightarrow 19.78 \pm 2.0023 \Rightarrow (17.78,\ 21.78)$.

13.39 **a**   The 95% prediction interval for an observation to be made when x* = 40 would be
$6.5511 \pm 2.15\sqrt{(0.0356)^2 + (0.008955)^2} \Rightarrow 6.5511 \pm 2.15(0.0367)$
$\Rightarrow 6.5511 \pm 0.0789 = (6.4722,\ 6.6300)$.

b   The 99% prediction interval for an observation to be made when x* = 35 would be
$6.5876 \pm 2.98\sqrt{(0.0356)^2 + (0.009414)^2} \Rightarrow 6.5876 \pm 2.98(0.0368)$
$\Rightarrow 6.5876 \pm 0.1097 = (6.4779,\ 6.6973)$.

c   Yes, because x* = 60 is farther from the mean value of x, which is 42.375, than is 40 or 35.

13.41 **a**   From the MINITAB output we get

Clutch Size = −133.02 + 5.92 Snout-Vent Length

b   From the MINITAB output we get $s_b = 1.127$

c   Let β denote the mean increase in Clutch size associated with a one unit increase in Snout-Vent length.

$H_o: \beta = 0$ $H_a: \beta > 0$

$\alpha = 0.05$ (a significance level is not specified in the problem so we use 0.05 for illustration)

The test statistic is: $t = \dfrac{b}{s_b}$ with df. = 12.

From the MINITAB output $t = \dfrac{5.919}{1.127} = 5.25$

P-value = area under the 12 df t curve to the right of 5.25 $\approx$ 0.

Hence the null hypothesis is rejected. The data provide strong evidence indicating that the slope is positive.

d The predicted value of the clutch size for a salamander with snout-vent length of 65 is $-133.02 + 5.919(65) = 251.715$.

e The value 205 is very much outside the range of snout-vent length values in the available data. The validity of the estimated regression line this far away from the range of x values in the data set is highly questionable. Therefore, calculation of a predicted value and/or a prediction interval for the clutch size for a salamander with snout-vent length of 205 based on available data is not recommended.

13.43 **a**

$$b = \frac{1081.5 - \left[\dfrac{(269)(51)}{14}\right]}{7445 - \left[\dfrac{(269)^2}{14}\right]} = \frac{101.571}{2276.357} = 0.04462$$

$a = 3.6429 - (0.04462)(19.214) = 2.78551$

The equation of the estimated regression line is $\hat{y} = 2.78551 + 0.04462x$.

b $H_o: \beta = 0 \quad H_a: \beta \neq 0$

$\alpha = 0.05$

The test statistic is: $t = \dfrac{b}{s_b}$ with df. = 12.

SSResid = $190.78 - (2.78551)(51) - (0.00462)(1081.5) = 0.46246$

$s_e^2 = \dfrac{0.46246}{12} = 0.0385$

$s_b^2 = \dfrac{0.0385}{2276.357} = 0.0000169, \quad s_b = 0.004113$

$t = \dfrac{0.04462}{0.004113} = 10.85$

P-value = 2(area under the 12 df t curve to the right of 10.85) = 2(0) = 0.

Since the P-value is less than α, the null hypothesis is rejected. The data suggests that the simple linear regression model provides useful information for predicting moisture content from knowledge of time.

c The point estimate of the moisture content of an individual box that has been on the shelf 30 days is $2.78551 + 0.04462(30) = 4.124$.

The 95% prediction interval is

$$4.124 \pm (2.18)\sqrt{0.0385}\sqrt{1 + \frac{1}{14} + \frac{(30-19.214)^2}{2276.357}}$$

$$\Rightarrow 4.124 \pm 2.18(0.2079) \Rightarrow 4.124 \pm 0.453 = (3.671, 4.577).$$

d Since values greater than equal to 4.1 are included in the interval constructed in **c**, it is very plausible that a box of cereal that has been on the shelf 30 days will not be acceptable.

13.45 a

$$b = \frac{57760 - \left[\dfrac{(1350)(600)}{15}\right]}{155400 - \left[\dfrac{(1350)^2}{15}\right]} = \frac{3760}{33900} = 0.1109$$

$a = 40 - (0.1109)(90) = 30.019$

The equation for the estimated regression line is $\hat{y} = 30.019 + 0.1109x$.

b When $x = 100$, the point estimate of $\alpha + \beta(100)$ is $30.019 + 0.1109(100) = 41.109$.

SSResid $= 24869.33 - (30.019)(600) - (0.1109)(57760) = 452.346$

$$s_e^2 = \frac{452.346}{13} = 34.7958$$

$$s_{a+b(100)}^2 = 34.7958\left[\frac{1}{15} + \frac{(100-90)^2}{33900}\right] = 2.422$$

$$s_{a+b(100)} = \sqrt{2.422} = 1.5564$$

The 90% confidence interval for the mean blood level for people who work where the air lead level is 100 is $41.109 \pm (1.77)(1.5564) \Rightarrow 41.109 \pm 2.755 \Rightarrow (38.354, 43.864)$.

c The prediction interval is $41.109 \pm (1.77)\sqrt{34.7958 + 2.422} \Rightarrow 41.109 \pm 10.798$ $\Rightarrow (30.311, 51.907)$.

d The interval of part **b** is for the mean blood level of all people who work where the air lead level is 100. The interval of part **c** is for a single randomly selected individual who works where the air lead level is 100.

13.47   **a**   The 95% prediction interval for sunburn index when distance is 35 is

$$2.5225 \pm 2.16 \sqrt{(0.25465)^2 + (0.07026)^2} \Rightarrow 2.5225 \pm 0.5706 \Rightarrow (1.9519, 3.0931).$$

The 95% prediction interval for sunburn index when distance is 45 is

$$1.9575 \pm 2.16 \sqrt{(0.25465)^2 + (0.06857)^2} \Rightarrow 1.9575 \pm 0.5696 \Rightarrow (1.3879, 2.5271).$$

The pair of intervals form a set of simultaneous prediction intervals with prediction level of at least $[100 - 2(5)]\% = 90\%$.

     b   The simultaneous prediction level would be at least $[100 - 3(1)]\% = 97\%$.

Section 13.5

13.49   The quantity r is a statistic as its value is calculated from the sample. It is a measure of how strongly the sample x and y values are linearly related. The value of r is an estimate of ρ. The quantity ρ is a population characteristic. It measures the strength of association between the x and y values in the population.

13.51   Let ρ denote the true correlation coefficient between teaching evaluation index and annual raise.

$H_o: \rho = 0 \qquad H_a: \rho \neq 0$

$\alpha = 0.05$

$$t = \frac{r}{\sqrt{\dfrac{1-r^2}{n-2}}} \qquad \text{with df.} = 351$$

$n = 353, \ r = 0.11$

$$t = \frac{0.11}{\sqrt{\dfrac{1-(0.11)^2}{351}}} = \frac{0.11}{0.05305} = 2.07$$

The t curve with 351 df is essentially the z curve.

P-value = 2(area under the z curve to the right of 2.07) = 2(0.0192) = 0.0384.

Since the P-value is less than α, H_o is rejected. There is sufficient evidence in the sample to conclude that there appears to be a linear association between teaching evaluation index and annual raise.

According to the guidelines given in the text book, $r = 0.11$ suggests only a weak linear relationship. Since $r^2 = 0.0121$, fitting the simple linear regression model to the data would result in only about 1.21% of observed variation in annual raise being explained.

13.53   **a**   Let ρ denote the correlation coefficient between time spent watching television and grade point average in the population from which the observations were selected.

$H_o: \rho = 0 \qquad H_a: \rho < 0$

$\alpha = 0.01$

$$t = \frac{r}{\sqrt{\dfrac{1-r^2}{n-2}}} \qquad \text{with df.} = 526$$

$n = 528, \ r = -0.26$

$$t = \frac{-0.26}{\sqrt{\dfrac{1-(-0.26)^2}{526}}} = \frac{-0.26}{0.042103} = -6.175$$

The t curve with 526 df is essentially the z curve.

P-value = area under the z curve to the left $-6.175 \approx 0$.

Since the P-value is less than α, H_o is rejected. The data does support the conclusion that there is a negative correlation in the population between the two variables, time spent watching television and grade point average.

b   The coefficient of determination measures the proportion of observed variation in grade point average explained by the regression on time spent watching television. This value would be $(-0.26)^2 = 0.0676$. Thus only 6.76% of the observed variation in grade point average would be explained by the regression. This is not a substantial percentage.

13.55   From the summary quantities:

$$S_{xy} = 673.65 - \left[\frac{(136.02)(39.35)}{9} \right] = 78.94$$

$$S_{xx} = 3602.65 - \left[\frac{(136.02)^2}{9} \right] = 1546.93$$

$$S_{yy} = 184.27 - \left[\frac{(39.35)^2}{9} \right] = 12.223$$

$$r = \frac{78.94}{\sqrt{(1546.93)(12.223)}} = \frac{78.94}{137.51} = 0.574$$

Let ρ denote the correlation between surface and subsurface concentration.

$H_o: \rho = 0$ $H_a: \rho \neq 0$

$\alpha = 0.05$

$$t = \frac{r}{\sqrt{\frac{1-r^2}{n-2}}} \quad \text{with df.} = 7$$

$$t = \frac{0.574}{\sqrt{\frac{1-(0.574)^2}{7}}} = 1.855$$

P-value = 2(area under the 7 df t curve to the right of 1.855) $\approx$ 2(0.053) = 0.106.

Since the P-value exceeds α, H_o is not rejected. The data does not support the conclusion that there is a linear relationship between surface and subsurface concentration.

13.57 $H_o: \rho = 0$ $H_a: \rho \neq 0$

$\alpha = 0.05$

$$t = \frac{r}{\sqrt{\frac{1-r^2}{n-2}}} \quad \text{with df.} = 9998$$

$n = 10000, \ r = 0.022$

$$t = \frac{0.022}{\sqrt{\frac{1-(0.022)^2}{9998}}} = 2.2$$

The t curve with 9998 df is essentially the z curve.

P-value = 2(area under the z curve to the right of 2.2) = 2(0.0139) = 0.0278.

Since the P-value is less than α, H_o is rejected. The results are statistically significant. Because of the extremely large sample size, it is easy to detect a value of ρ which differs from zero by a small amount. If ρ is very close to zero, but not zero, the practical significance of a non-zero correlation may be of little consequence.

Supplementary Exercises

13.59 **a** $t = \dfrac{-0.18}{\sqrt{\dfrac{1-(-.18)^2}{345}}} = \dfrac{-0.18}{0.052959} = -3.40$ with df = 345

If the test was a one-sided test, then the P-value equals the area under the z curve to the left of −3.40, which is equal to 0.0003. It the test was a two-sided test, then the P-value is 2(0.0003) = 0.0006. While the researchers' statement is true, they could have been more precise in their statement about the P-value.

b From my limited experience, I have observed that the more visible a person's sense of humor, the less depressed they *appear* to be. This would suggest a negative correlation between Coping Humor Scale and Sense of Humor.

c Since $r^2 = (-0.18)^2 = 0.0324$, only about 3.24% of the observed variability in sense of humor can be explained by the linear regression model. This suggests that a simple linear regression model may not give accurate predictions.

13.61 **a** When x* = 20, a+b(20) = 2.2255 + 0.1521(20) = 5.2675.

$$s^2_{a+b(20)} = 0.1187\left[\frac{1}{11} + \frac{(20-26.627)^2}{342.622}\right] = 0.026$$

$$s_{a+b(20)} = \sqrt{0.026} = 0.1613$$

The 95% confidence interval for the mean tensile modulus when bound rubber content is 20% is $5.2675 \pm 2.26(0.1613) \Rightarrow 5.2675 \pm 0.3645 \Rightarrow (4.903, 5.632)$

b From **a**, $\hat{y} = 5.2675$ and the 95% prediction interval is

$$5.2675 \pm 2.26\sqrt{0.1187 + (0.1613)^2} \Rightarrow 5.2675 \pm 0.8597 \Rightarrow (4.4078, 6.1272)$$

c The request is to estimate the "true <u>mean</u> tensile strength" at a given value of x, not to estimate the "<u>change</u> in mean tensile strength" associated with a one-unit increase in x.

13.63 **a** $H_o: \beta = 0$ $H_a: \beta \neq 0$

$\alpha = 0.01$

$t = \dfrac{b}{s_b}$

From the Minitab output, t = −3.95 and the P-value = 0.003. Since the P-value is less than α, H_o is rejected. The data supports the conclusion that the simple linear regression model is useful.

b   A 95% confidence interval for β is $-2.3335 \pm 2.26(0.5911) \Rightarrow -2.3335 \pm 1.3359$
$\Rightarrow (-3.6694, -0.9976)$.

c $a+b(10) = 88.796 - 2.3335(10) = 65.461$

$s_{a+b(10)} = 0.689$

The 95% prediction interval for an individual y when x = 10 is

$$65.461 \pm 2.26\sqrt{(0.689)^2 + 4.789} \Rightarrow 65.461 \pm 5.185 \Rightarrow (60.276, 70.646).$$

d   Because x = 11 is farther from $\overline{x}$ than x = 10 is from $\overline{x}$.

13.65 a   Let ρ denote the correlation coefficient between soil hardness and trail length.

$H_o: \rho = 0 \quad H_a: \rho < 0$

$\alpha = 0.05$

$$t = \frac{r}{\sqrt{\dfrac{1-r^2}{(n-2)}}} \quad \text{with df.} = 59$$

$$t = \frac{-0.6213}{\sqrt{\dfrac{1-(-0.6213)^2}{59}}} = -6.09$$

P-value = area under the 59 df t curve to the left of $-6.09 \approx 0$.

Since the P-value is less than α, the null hypothesis is rejected. The data supports the conclusion of a negative correlation between trail length and soil hardness.

b   When x* = 6, a+b(6) = 11.607 – 1.4187(6) = 3.0948

$$s_{a+b(6)}^2 = (2.35)^2 \left[\frac{1}{61} + \frac{(6-4.5)^2}{250} \right] = 0.1402$$

$$s_{a+b(6)} = \sqrt{0.1402} = 0.3744$$

The 95% confidence interval for the mean trail length when soil hardness is 6 is

$$3.0948 \pm 2.00(0.3744) \Rightarrow 3.0948 \pm 0.7488 \Rightarrow (2.346, 3.844).$$

c   When x* = 10, a+b(10) = 11.607 – 1.4187(10) = –2.58

According to the least-squares line, the predicted trail length when soil hardness is 10 is −2.58. Since trail length cannot be negative, the predicted value makes no sense. Therefore, one would not use the simple linear regression model to predict trail length when hardness is 10.

13.67 $n = 17$, $\Sigma x = 821$, $\Sigma x^2 = 43447$, $\Sigma y = 873$, $\Sigma y^2 = 46273$, $\Sigma xy = 40465$,

$$S_{xy} = 40465 - \left[\frac{(821)(873)}{17}\right] = 40465 - 42160.7647 = -1695.7647$$

$$S_{xx} = 43447 - \left[\frac{(821)^2}{17}\right] = 43447 - 39649.4706 = 3797.5294$$

$$S_{yy} = 46273 - \left[\frac{(873)^2}{17}\right] = 46273 - 44831.1176 = 1441.8824$$

$$b = \frac{-1695.7647}{3797.5294} = -0.4465$$

$$a = 51.3529 - (-0.4465)(48.2941) = 72.9162$$

$$\text{SSResid} = 46273 - 72.9162(873) - (-0.4465)(40465) = 684.78$$

$$s_e^2 = \frac{684.78}{15} = 45.652 \ , \quad s_e = 6.7566$$

$$s_b = \frac{6.7566}{\sqrt{3797.5294}} = 0.1096$$

a   Let β denote the average change in percentage area associated with a one year increase in age.

$H_o: \beta = -0.5 \quad H_a: \beta \neq -0.5$

$\alpha = 0.10$

$$t = \frac{b - (-0.5)}{s_b} \quad \text{with df.} = 15$$

$$t = \frac{-0.4465 - (-0.5)}{0.1096} = 0.49$$

P-value = 2(area under the 15 df t curve to the right of 0.49) ≈ 2(0.312) = 0.624.

Since the P-value exceeds α, H_0 is not rejected. There is not sufficient evidence in the sample to contradict the prior belief of the researchers.

b   When $x^* = 50$, $\hat{y} = 72.9162 + (-0.4465)(50) = 50.591$

$$s_{a+b(50)} = 6.7471\sqrt{\frac{1}{17} + \frac{(50 - 48.2941)^2}{3797.5294}} = 1.647$$

The 95% confidence interval for the true average percent area covered by pores for all 50 year-olds is

$$50.591 \pm (2.13)(1.647) \Rightarrow 50.591 \pm 3.508 \Rightarrow (47.083, 54.099).$$

13.69   The 95% confidence interval for α is

$$20.1251 \pm 2.23(0.9402) \Rightarrow 20.1251 \pm 2.0967 \Rightarrow (18.0284, 22.2218).$$

13.71   For data set 1:

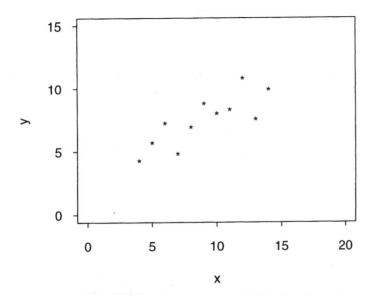

The plot above supports the appropriateness of fitting a simple linear regression model to data set 1.

For data set 2:

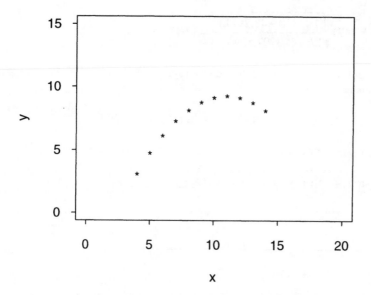

This plot suggests quite clearly that the fitting of a simple linear regression model to data set 2 would not be appropriate.

For data set 3:

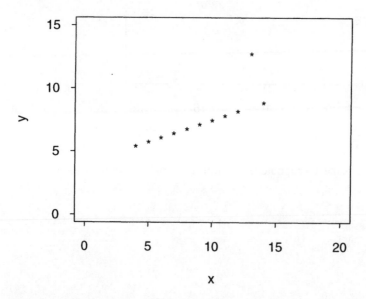

This plot reveals an observation which would have an unusually large residual. A simple linear regression model would not be appropriate for this data set, but might be for the data set with the one unusual observation deleted.

For data set 4:

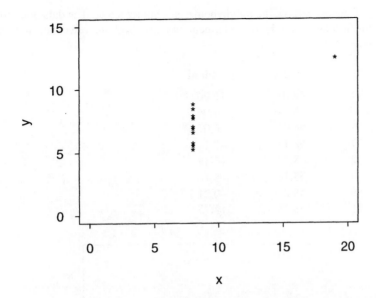

This plot reveals one point that is a very influential point. In fact, the slope is determined solely by this observation. The simple linear regression model would not be appropriate for this data set.

13.73 Summary values are: n = 8,

$S_{xx} = 42$, $S_{yy} = 586.875$, $S_{xy} = 1.5$

a $b = \dfrac{1.5}{42} = 0.0357$

a = 58.125 − (0.0357)(4.5) = 57.964

The equation of the estimated regression line is $\hat{y} = 57.964 + 0.0357x$.

b   Let β denote the expected change in glucose concentration associated with a one day increase in fermentation time.

H_o: β = 0 H_a: β ≠ 0

α = 0.10

The test statistic: $t = \dfrac{b}{s_b}$ with df. = 6.

From the data, $s_b = 1.526$ and $t = \dfrac{0.0357}{1.526} = 0.023$.

13-24

P-value = 2(area under the 6 df t curve to the right of 0.023) = 0.982.

Since the P-value exceeds α, the null hypothesis is not rejected. The data does not indicate a linear relationship between fermentation time and glucose concentration.

c

x	y	Pred-y	Residual
1	74	58.00	16.00
2	54	58.04	−4.04
3	52	58.07	−6.07
4	51	58.11	−7.11
5	52	58.14	−6.14
6	53	58.18	−5.18
7	58	58.21	−0.21
8	71	58.25	12.75

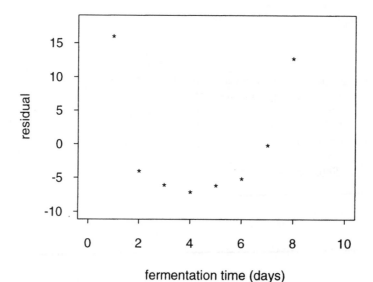

fermentation time (days)

d   The residual plot has a very distinct curvilinear pattern which indicates that a simple linear regression model is not appropriate for describing the relationship between y and x. Instead, a model incorporating curvature should be fit.

13.75   **a**   Summary values are: n = 17, S_{xx} = 13259.0706, S_{yy} = 1766.4706, S_{xy} = −79.9294, where x = depth and y = zinc concentration.

$$r = \frac{-79.9294}{\sqrt{(13259.0706)(1766.4706)}} = -0.0165$$

Let ρ denote the correlation between depth and zinc concentration.

$H_o: \rho = 0 \quad H_a: \rho \neq 0$

$\alpha = 0.05$

$$t = \frac{r}{\sqrt{\dfrac{1-r^2}{(n-2)}}} \quad \text{with df} = 15.$$

$$t = \frac{(-0.0165)}{\sqrt{\dfrac{1-(-0.0165)^2}{15}}} = -0.06$$

P-value = 2(area under the 15 df t curve to the left of -0.06) $\approx 2(0.47) = 0.94$.

Since the P-value exceeds α, the null hypothesis is not rejected. The data suggests that no correlation exists between depth and zinc concentration.

b   Summary values are: n = 17, $\Sigma x = 531.7$, $\Sigma x^2 = 29{,}888.77$, $\Sigma y = 58.5$,
$\Sigma y^2 = 204.51$, $\Sigma xy = 1983.64$, $S_{xx} = 13259.0706$, $S_{yy} = 3.2012$, $S_{xy} = 153.9665$,
where x = depth and y = iron concentration.

$$r = \frac{153.9665}{\sqrt{(13259.30706)(3.2012)}} = 0.747$$

Let ρ denote the correlation between depth and iron concentration.

$H_o: \rho = 0 \quad H_a: \rho \neq 0$

$\alpha = 0.05$

$$t = \frac{r}{\sqrt{\dfrac{1-r^2}{(n-2)}}} \quad \text{with df.} = 15$$

$$t = \frac{0.747}{\sqrt{\dfrac{1-(0.747)^2}{15}}} = 4.35$$

P-value = 2(area under the 15 df t curve to the right of 4.35) $\approx 2(0.0003) = 0.0006$.

Since the P-value is less than α, the null hypothesis is rejected. The data does suggest a correlation between depth and iron concentration.

c $$b = \frac{153.9665}{13259.0706} = 0.01161$$

13-26

$$a = \frac{58.5 - (0.01161)(531.7)}{17} = 3.0781$$

d   When $x^* = 50$, $a + b(50) = 3.6586$.

$$SSResid = 204.51 - (3.0781)(58.5) - (0.01161)(1983.64) = 1.411$$

$$s_e^2 = \frac{1.411}{15} = 0.0941$$

$$s_{a+b(50)}^2 = 0.0941\left[\frac{1}{17} + \frac{(50 - 31.276)^2}{13259.0706}\right] = 0.00802$$

The 95% prediction interval for the iron concentration of a single core sample taken at a depth of 50m. is

$$3.6586 \pm (2.13)\sqrt{0.0941 + 0.00802} \Rightarrow 3.6586 \pm (2.13)(0.3196)$$
$$\Rightarrow 3.6585 \pm 0.6807 \Rightarrow (2.9778, 4.3392).$$

e   When $x^* = 70$, $a + b(70) = 3.8908$

$$s_{a+b(70)}^2 = 0.0941\left[\frac{1}{17} + \frac{(70 - 31.276)^2}{13259.0706}\right] = 0.0162$$

$$s_{a+b(70)} = \sqrt{0.0162} = 0.1273$$

The 95% confidence interval for $\alpha + \beta(70)$ is

$$3.8908 \pm (2.13)(0.1273) \Rightarrow 3.8908 \pm 0.2711 \Rightarrow (3.6197, 4.1619).$$

With 95% confidence it is estimated that the mean iron concentration at a depth of 70m is between 3.6197 and 4.1619.

13.77   **a**   The e_i's are the deviations of the observations from the population regression line, whereas the residuals are the deviations of the observations from the estimated regression line.

        b   The simple linear regression model states that $y = \alpha + \beta x + e$. Without the random deviation e, the equation implies a deterministic model, whereas the simple linear regression model is probabilistic.

        c   The quantity b is a statistic. Its value is known once the sample has been collected, and different samples result in different b values. Therefore, it does not make sense to test hypotheses about b. Only hypotheses about a population characteristic can be tested.

d   If r = +1 or −1, then each point falls exactly on the regression line and SSResid would equal zero. A true statement is that SSResid is always greater than or equal to zero.

e   The sum of the residuals must equal zero. Thus, if they are not all exactly zero, at least one must be positive and at least one must be negative. They cannot all be positive. Since there are some positive and no negative values among the reported residuals, the student must have made an error.

f $\text{SSTo} = \sum (y - \bar{y})^2$ must be greater than or equal to $\text{SSResid} = \sum (y - \hat{y})^2$. Thus, the values given must be incorrect.

Chapter 14
Multiple Regression Analysis

Section 14.1

14.1   A deterministic model does not have the random deviation component e, while a probabilistic model does contain such a component.

Let   y = total number of goods purchased at a service station which sells only one grade
            of gas and one type of motor oil.
        x_1 = gallons of gasoline purchased
        x_2 = number of quarts of motor oil purchased.

Then y is related to x_1 and x_2 in a deterministic fashion.

Let   y = IQ of a child
        x_1 = age of the child
        x_2 = total years of education of the parents.

Then y is related to x_1 and x_2 in a probabilistic fashion.

14.3   The following multiple regression model is suggested by the given statement.
$$y = \beta_0 + \beta_1 x_1 + \beta_2 x_2 + e .$$

An interaction term is not included in the model because it is given that x_1 and x_2 make independent contributions to academic achievement.

14.5   **a** $415.11 - 6.6(20) - 4.5(40) = 103.11$

b $415.11 - 6.6(18.9) - 4.5(43) = 96.87$

c $\beta_1 = -6.6$. Hence 6.6 is the expected *decrease* in yield associated with a one-unit increase in mean temperature between date of coming into hop and date of picking when the mean percentage of sunshine remains fixed.

$\beta_2 = -4.5$. So 4.5 is the expected *decrease* in yield associated with a one-unit increase in mean percentage of sunshine when mean temperature remains fixed.

14.7   **a**

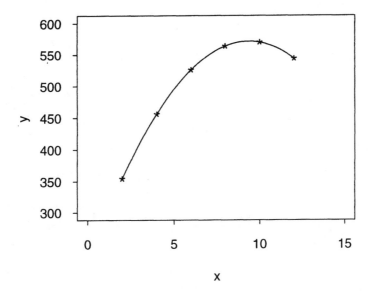

b   The mean chlorine content at x = 8 is 564, while at x = 10 it is 570. So the mean chlorine content is higher for x = 10 than for x = 8.

c   When x = 9, y = 220 + 75(9) − 4(9)^2 = 571.
The change in mean chlorine content when the degree of delignification increases from 8 to 9 is 571 − 564 = 7. The change in mean chlorine content when the degree of delignification increases from 9 to 10 is 570 − 571 = −1.

14.9   **a**

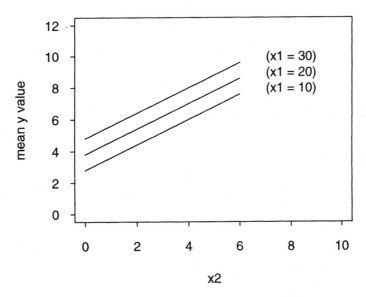

For x$_1$ = 30,   y = 4.8 + 0.8x$_2$; for x$_1$ = 20,   y = 3.8 + 0.8x$_2$; for x$_1$ = 10,   y = 2.8 + 0.8x$_2$

b

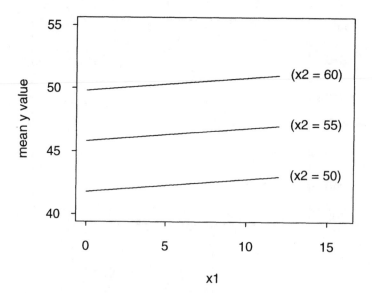

For $x_2 = 60$, $y = 49.8 + 0.1x_2$; for $x_2 = 55$, $y = 45.8 + 0.1x_2$; for $x_2 = 50$, $y = 41.8 + 0.1x_2$

c   The parallel lines in each graph are attributable to the lack of interaction between the two independent variables.

d

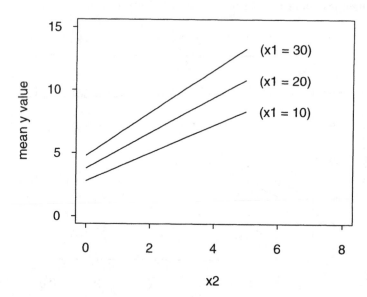

For $x_1 = 30$, $y = 4.8 + 1.7x_2$; for $x_1 = 20$, $y = 3.8 + 1.4x_2$; for $x_1 = 10$, $y = 2.8 + 1.1x_2$

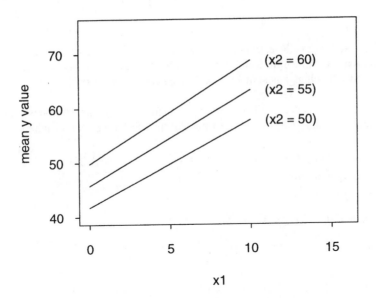

For $x_2 = 60$, $y = 49.8 + 1.9x_2$; for $x_2 = 55$, $y = 45.8 + 1.75x_2$;
for $x_2 = 50$, $y = 41.8 + 1.6x_2$

Because there is an interaction term, the lines are not parallel.

14.11 **a** $y = \alpha + \beta_1x_1 + \beta_2x_2 + \beta_3x_3 + e$

b $y = \alpha + \beta_1x_1 + \beta_2x_2 + \beta_3x_3 + \beta_4x_1^2 + \beta_5x_2^2 + \beta_6x_3^2 + e$

c $y = \alpha + \beta_1x_1 + \beta_2x_2 + \beta_3x_3 + \beta_4x_1x_2 + e$
$y = \alpha + \beta_1x_1 + \beta_2x_2 + \beta_3x_3 + \beta_4x_1x_3 + e$
$y = \alpha + \beta_1x_1 + \beta_2x_2 + \beta_3x_3 + \beta_4x_2x_3 + e$

d $y = \alpha + \beta_1x_1 + \beta_2x_2 + \beta_3x_3 + \beta_4x_1^2 + \beta_5x_2^2 + \beta_6x_3^2 + \beta_7x_1x_2 + \beta_8x_1x_3 + \beta_9x_2x_3 + e$

14.13 **a** Three dummy variables would be needed to incorporate a non-numerical variable with four categories.

$x_3 = 1$ if the car is a sub-compact, 0 otherwise

$x_4 = 1$ if the car is a compact, 0 otherwise

$x_5 = 1$ if the car is a midsize, 0 otherwise

$y = \alpha + \beta_1x_1 + \beta_2x_2 + \beta_3x_3 + \beta_4x_4 + \beta_5x_5 + e$

b $x_6 = x_1x_3$, $x_7 = x_1x_4$, and $x_8 = x_1x_5$ are the additional predictors needed to incorporate interaction between age and size class.

Section 14.2

14.15 **a** $b_3 = -0.0096$ is the estimated change in mean VO_2 max associated with a one-unit increase in "one mile walk time" when the value of the other predictor variables are fixed. The change is actually a decrease since the regression coefficient is negative.

b $b_1 = 0.6566$ is the estimated difference in mean VO_2 max for males versus females (male minus female) when the values of the other predictor variables are fixed.

c
$$\hat{y} = 3.5959 + .6566\,(1) + .0096\,(80) - .0996\,(11) - .0080\,(140)$$
$$= 3.5959 + .6566 + .7680 - 1.0956 - 1.12$$
$$= 2.8049$$
$$\text{residual} = y - \hat{y} = 3.15 - 2.8049 = 0.3451$$

d
$$R^2 = 1 - \frac{SSResid}{SSTo} = 1 - \frac{30.1033}{102.3922} = 1 - .294 = .706$$

e
$$s_e^2 = \frac{SSResid}{n - (k+1)} = \frac{30.1033}{196 - 5} = \frac{30.1033}{191} = .157609$$

$$s_e = \sqrt{.157609} = .397$$

14.17 **a** $0.05 > \text{P-value} > 0.01$

b $\text{P-value} > 0.10$

c $\text{P-value} = 0.01$

d $0.01 > \text{P-value} > 0.001$

14.19 The fitted model was $y = \alpha + \beta_1 x_1 + \beta_2 x_2 + \beta_3 x_3 + \beta_4 x_4 + \beta_5 x_5 + \beta_6 x_6 + e$

$H_o: \beta_1 = \beta_2 = \beta_3 = \beta_4 = \beta_5 = \beta_6 = 0$

$H_a:$ at least one among $\beta_1, \beta_2, \beta_3, \beta_4, \beta_5, \beta_6$ is not zero

$\alpha = 0.01$

$$F = \frac{R^2/k}{(1 - R^2)/[n - (k+1)]}$$

$n = 37, \ df_1 = k = 6, \ df_2 = n - (k+1) = 37 - 7 = 30$

$R^2 = 0.83$

$$F = \frac{R^2/k}{(1-R^2)/[n-(k+1)]} = \frac{.83/6}{(1-.83)/30} = 24.41$$

From Appendix Table VII, 0.001 > P-value.

Since the P-value is less than α, the null hypothesis is rejected. There does appear to be a useful linear relationship between y and at least one of the six predictors.

14.21   The fitted model was $y = \alpha + \beta_1 x_1 + \beta_2 x_2 + \beta_3 x_3 + \beta_4 x_4 + e$

$H_o: \beta_1 = \beta_2 = \beta_3 = \beta_4 = 0$

H_a: at least one among $\beta_1, \beta_2, \beta_3, \beta_4$, is not zero

$\alpha = 0.05$

$$F = \frac{R^2/k}{(1-R^2)/[n-(k+1)]}$$

$n = 196$, $df_1 = k = 4$, $df_2 = n - (k+1) = 196 - 5 = 191$

$R^2 = 0.706$

$$F = \frac{R^2/k}{(1-R^2)/[n-(k+1)]} = \frac{.706/4}{(1-.706)/191} = 114.66$$

From Appendix Table VII, 0.001 > P-value.

Since the P-value is less than α, the null hypothesis is rejected. There does appear to be a useful linear relationship between y and at least one of the four predictors.

14.23   **a**   Estimated mean value of $y = 86.85 - 0.12297x_1 + 5.090x_2 - 0.07092x_3 + 0.001538x_4$

   b $H_o: \beta_1 = \beta_2 = \beta_3 = \beta_4 = 0$

H_a: at least one among $\beta_1, \beta_2, \beta_3, \beta_4$, is not zero

$\alpha = 0.01$

$$F = \frac{R^2/k}{(1-R^2)/[n-(k+1)]}$$

$n = 31$, $df_1 = k = 4$, $df_2 = n - (k+1) = 31 - 5 = 26$

$R^2 = 0.908$

$$F = \frac{R^2/k}{(1-R^2)/[n-(k+1)]} = \frac{.908/4}{(1-.908)/26} = 64.15$$

From Appendix Table VII, $0.001 > \text{P-value}$

Since the P-value is less than α, the null hypothesis is rejected. There does appear to be a useful linear relationship between y and at least one of the four predictors.

c $R^2 = 0.908$. This means that 90.8% of the variation in the observed tar content values has been explained by the fitted model.

$s_e = 4.784$. This means that the typical distance of an observation from the corresponding mean value is 4.784.

14.25 a The regression model being fitted is $y = \alpha + \beta_1 x_1 + \beta_2 x_2 + e$.
Using MINITAB, the regression command yields the following output.

```
The regression equation is
weight(g) = - 511 + 3.06 length(mm) - 1.11 age(years)

Predictor          Coef        SE Coef            T          P
Constant         -510.9          286.1        -1.79      0.096
length(mm)       3.0633         0.8254         3.71      0.002
age(years)       -1.113          9.040        -0.12      0.904

S = 94.24          R-Sq = 59.3%       R-Sq(adj) = 53.5%

Analysis of Variance

Source            DF           SS           MS         F          P
Regression         2       181364        90682     10.21      0.002
Residual Error    14       124331         8881
Total             16       305695
```

b $H_0: \beta_1 = \beta_2 = 0$

H_a: At least one of the two β_i's is not zero.

$$F = \frac{\text{SSRegr}/k}{\text{SSResid}/[n-(k+1)]}$$

$n = 17$, $df_1 = k = 2$, $df_2 = n - (k+1) = 17 - 3 = 14$

$$F = \frac{\text{SSRegr}/k}{\text{SSResid}/[n-(k+1)]} = \frac{181364/2}{124331/14} = 10.21$$

14-7

From the Minitab output, the P-value = 0.002. Since the P-value is less than 0.05 (we have chosen $\alpha = 0.05$ for illustration) the null hypothesis is rejected. The data suggests that the multiple regression model is useful for predicting weight.

c   Performing a best subsets regression in MINITAB yields the following output.

```
Response is y
                                                          x x
Vars    R-Sq    R-Sq(adj)        C-p           S        1 2

  1     59.3      56.6           1.0        91.092      X
  1     19.3      13.9          14.8       128.23         X
  2     59.3      53.5           3.0        94.238      X X
```

Among the two models with only a single predictor, we see that the model with x_1 alone is superior to the model with x_2 alone (the value of R-sq for the model with only x_1 is 0.593 whereas R-sq is equal to 0.193 for the model with x_2 only). Also, the model with x_1 as the only predictor has the smallest value for the Cp statistic. Hence we conclude that the predictor 'age' may be eliminated from the regression model without significantly affecting our ability to make accurate predictions.

d   Let $x_3 = 1$ if the year caught is 1995 and 0 otherwise. Using MINITAB to regress y on x_1, x_2, and x_3 yields the following output.

```
The regression equation is
y = - 1754 + 5.71 x1 + 2.48 x2 + 238 x3

Predictor          Coef       SE Coef          T          P
Constant         -1753.8        333.2       -5.26      0.000
x1                5.7149       0.7972        7.17      0.000
x2                2.479         5.925        0.42      0.683
x3              237.95         52.96         4.49      0.001

S = 61.20       R-Sq = 84.1%      R-Sq(adj) = 80.4%
```

The t-ratio for x_3 is 4.49 with a corresponding P-value = 0.001. Since the P-value is less than $\alpha = 0.05$, we conclude that year is a useful predictor even after length and age are included in the model.

14.27   a   Using MINITAB to fit the required regression model yields the following output.

```
The regression equation is
volume = - 859 + 23.7 minwidth + 226 maxwidth
              + 225 elongation

Predictor          Coef       SE Coef          T          P
Constant         -859.2        272.9       -3.15      0.005
minwidth          23.72        85.66        0.28      0.784
maxwidth         225.81        85.76        2.63      0.015
elongation       225.24        90.65        2.48      0.021
```

```
S = 287.0          R-Sq = 67.6%          R-Sq(adj) = 63.4%

Analysis of Variance

Source            DF        SS          MS         F       P
Regression         3     3960700     1320233     16.03   0.000
Residual Error    23     1894141       82354
Total             26     5854841
```

b   Adjusted R^2 takes into account the number of predictors used in the model whereas R^2 does not do so. In particular, adjusted R^2 enables us to make a "fair comparison" of the performances of models with differing numbers of predictors.

c $H_o: \beta_1 = \beta_2 = \beta_3 = 0$

H_a: At least one of the three β_i's is not zero.

$\alpha = 0.05$ (for illustration)

$$F = \frac{R^2/k}{(1-R^2)/[n-(k+1)]}$$

$n = 27$, $df_1 = k = 3$, $df_2 = n - (k + 1) = 27 - 4 = 23$

$R^2 = 0.676$

$$F = \frac{R^2/k}{(1-R^2)/[n-(k+1)]} = \frac{0.676/3}{(1-0.676)/23} = 16.03$$

The corresponding P-value is 0.000 (correct to 3 decimals). Since the P-value is less than α, the null hypothesis is rejected. There does appear to be a useful linear relationship between y and at least one of the three predictors.

14.29 **a**   SSResid = 390.4347

SSTo $= 7855.37 - 14(21.1071)^2 = 1618.2093$

SSRegr $= 1618.2093 - 390.4347 = 1227.7746$

b $R^2 = \dfrac{1227.7746}{1618.2093} = .759$

This means that 75.9 percent of the variation in the observed shear strength values has been explained by the fitted model.

c $H_o: \beta_1 = \beta_2 = \beta_3 = \beta_4 = \beta_5 = 0$

H_a: at least one among $\beta_1, \beta_2, \beta_3, \beta_4, \beta_5$, is not zero

$$\alpha = 0.05$$

$$F = \frac{R^2/k}{(1-R^2)/[n-(k+1)]}$$

$$n = 14, \ df_1 = k = 5, \ df_2 = n - (k+1) = 14 - 6 = 8$$

$$R^2 = 0.759$$

$$F = \frac{R^2/k}{(1-R^2)/[n-(k+1)]} = \frac{.759/5}{(1-.759)/8} = 5.039$$

From Appendix Table VII, $0.05 > $ P-value > 0.01.

Since the P-value is less than α, the null hypothesis is rejected. There does appear to be a useful linear relationship between y and at least one of the predictors. The data suggests that the independent variables as a group do provide information that is useful for predicting shear strength.

14.31 $H_0: \beta_1 = \beta_2 = 0$

H_a: At least one of the two β_i's is not zero.

$$\alpha = 0.01$$

$$F = \frac{R^2/k}{(1-R^2)/[n-(k+1)]}$$

$$n = 24, \ df_1 = k = 2, \ df_2 = n - (k+1) = 24 - 3 = 21$$

$$F = \frac{R^2/k}{(1-R^2)/[n-(k+1)]} = \frac{.902/2}{(1-.902)/21} = 96.64$$

From Appendix Table VII, $0.001 > $ P-value.

Since the P-value is less than α, the null hypothesis is rejected. The data suggests that the quadratic model does have utility for predicting yield.

14.33   Using MINITAB, the regression command yields the following output.

The regression equation is
Y = − 151 − 16.2 X1 + 13.5 X2 + 0.0935 X1-SQ − 0.253 X2-SQ + 0.0492 X1*X2.

Predictor	Coef	Stdev	t-ratio	p
Constant	−151.4	134.1	−1.13	0.292
X1	−16.216	8.831	−1.84	0.104
X2	13.476	8.187	1.65	0.138
X1-SQ	0.09353	0.07093	1.32	0.224
X2-SQ	−0.2528	0.1271	−1.99	0.082
X1*X2	0.4922	0.2281	2.16	0.063

It can be seen (except for differences due to rounding errors) that the estimated regression equation given in the problem is correct.

14.35 Using MINITAB, the regression command yields the following output.

The regression equation is INF_RATE = 35.8 − 0.68 AVE_TEMP + 1.28 AVE_RH

Predictor	Coef	Stdev	t-ratio	p
Constant	35.83	53.54	0.67	0.508
AVE_TEMP	−0.676	1.436	−0.47	0.641
AVE_RH	1.2811	0.4243	3.02	0.005

s = 22.98   R-sq = 55.0%   R-sq(adj) = 52.1%

Analysis of Variance

SOURCE	DF	SS	MS	F	p
Regression	2	20008	10004	18.95	0.000
Error	31	16369	528		
Total	33	36377			

H_o: $\beta_1 = \beta_2 = 0$

H_a: At least one of the two β_i's is not zero.

$$F = \frac{SSRegr/k}{SSResid/[n-(k+1)]}$$

$n = 34$, $df_1 = k = 2$, $df_2 = n - (k + 1) = 34 - 3 = 31$

$$F = \frac{SSRegr/k}{SSResid/[n-(k+1)]} = \frac{20008/2}{16369/31} = 18.95$$

From the Minitab output, the P-value ≈ 0. Since the P-value is less than α, the null hypothesis is rejected. The data suggests that the multiple regression model has utility for predicting infestation rate.

Section 14.3

14.37 **a** The degrees of freedom for error is $100 - (7 + 1) = 92$. From Appendix Table III, the critical t value is approximately 1.99.

The 95% confidence interval for β_3 is
$-0.489 \pm (1.99)(0.1044) \Rightarrow -4.89 \pm 0.208 \Rightarrow (-0.697, -0.281)$.

With 95% confidence, the change in the mean value of a vacant lot associated with a one unit increase in distance from the city's major east-west thoroughfare is a decrease of as little as 0.281 or as much as 0.697.

b $H_o: \beta_1 = 0 \quad H_a: \beta_1 \neq 0$

$\alpha = 0.05$

$t = \dfrac{b_1}{s_{b_1}}$ with df. = 92

$t = \dfrac{-.183}{.3055} = -0.599$

P-value = 2(area under the 92 df t curve to the left of -0.599) $\approx 2(0.275) = 0.550$.

Since the P-value exceeds α, the null hypotheses is not rejected. This means that there is not sufficient evidence to conclude that there is a difference in the mean value of vacant lots that are zoned for residential use and those that are not zoned for residential use.

14.39 **a** $H_o: \beta_1 = \beta_2 = 0$

H_a: At least one of the two β_i's is not zero.

$\alpha = 0.05$

$$F = \frac{R^2/k}{(1-R^2)/[n-(k+1)]}$$

$n = 50, \; df_1 = k = 2, \; df_2 = n - (k + 1) = 50 - 3 = 47, \; R^2 = 0.86$

$$F = \frac{R^2/k}{(1-R^2)/[n-(k+1)]} = \frac{.86/2}{(1-.86)/47} = 144.36$$

From Appendix Table VII, $0.001 >$ P-value.
Since the P-value is less than α, the null hypothesis is rejected. The data suggests that the quadratic regression model has utility for predicting MDH activity.

b $H_o: \beta_2 = 0$ $H_a: \beta_2 \neq 0$

$\alpha = 0.01$

$t = \dfrac{b_2}{s_{b_2}}$ with df. = 47

$t = \dfrac{.0446}{.0103} = 4.33$

P-value = 2(area under the 47 df t curve to the right of 4.33) $\approx 2(0) = 0$.

Since the P-value is less than α, the null hypothesis is rejected. The quadratic term is an important term in this model.

c   The point estimate of the mean value of MDH activity for an electrical conductivity level of 40 is

$-0.1838 + 0.0272(40) + 0.0446(40^2) = -0.1838 + 0.0272(40) + 0.0446(1600)$
$= 72.2642$.

The 90% confidence interval for the mean value of MDH activity for an electrical conductivity level of 40 is

$72.2642 \pm (1.68)(0.120) \Rightarrow 72.2642 \pm 0.2016 \Rightarrow (72.0626, 72.4658)$

14.41   **a**   The value 0.469 is an estimate of the expected change (increase) in the mean score of students associated with a one unit increase in the student's expected score holding time spent studying and student's grade point average constant.

b $H_o: \beta_1 = \beta_2 = \beta_3 = 0$

H_a: At least one of the three β_i's is not zero.

$\alpha = 0.05$

$F = \dfrac{R^2/k}{(1 - R^2)/[n - (k+1)]}$

$n = 107,\ df_1 = k = 3,\ df_2 = n - (k+1) = 107 - 4 = 103,\ R^2 = 0.686$

$F = \dfrac{R^2/k}{(1 - R^2)/[n - (k+1)]} = \dfrac{.686/3}{(1 - .686)/103} = 75.01$

From Appendix Table VII, 0.001 > P-value.

Since the P-value is less than α, the null hypothesis is rejected. The data suggests that there is a useful linear relationship between exam score and at least one of the three predictor variables.

c   The 95% confidence interval for β_2 is

$3.369 \pm (1.98)(0.456) \Rightarrow 3.369 \pm 0.903 \Rightarrow (2.466, 4.272)$.

d   The point prediction would be $2.178 + 0.469(75) + 3.369(8) + 3.054(2.8) = 72.856$.

e   The prediction interval would be $72.856 \pm (1.98) \sqrt{s_{\hat{e}}^2 + (1.2)^2}$.

To determine $s_{\hat{e}}^2$, proceed as follows. From the definition of R^2, it follows that SSResid $= (1-R^2)$SSTo. So SSResid $= (1-0.686)(10200) = 3202.8$.

Then, $s_{\hat{e}}^2 = \dfrac{3202.8}{103} = 31.095$.

The prediction interval becomes

$72.856 \pm (1.98) \sqrt{31.095 + (1.2)^2} \Rightarrow 72.856 \pm (1.98)(5.704)$
$\Rightarrow 72.856 \pm 11.294 \Rightarrow (61.562, 84.150)$.

14.43 $H_o: \beta_3 = 0$ $H_a: \beta_3 \neq 0$

$\alpha = 0.05$

$t = \dfrac{b_3}{s_{b_3}}$ with d.f. $= 363$

$t = \dfrac{.00002}{.000009} = 2.22$

P-value $= 2$(area under the 363 df t curve to the right of 2.22) $\approx 2(0.014) = 0.028$.

Since the P-value is less than α, the null hypothesis is rejected. The conclusion is that the inclusion of the interaction term is important.

14.45   a $H_o: \beta_1 = \beta_2 = \beta_3 = 0$

H_a: At least one of the three β_i's is not zero.

$\alpha = 0.05$

Test statistic: $F = \dfrac{\text{SSRegr}/k}{\text{SSResid}/[n-(k+1)]}$

$$F = \frac{5073.4/3}{1854.1/6} = 5.47$$

From Appendix Table VII, $0.05 > \text{P-value} > 0.01$.

Since the P-value is less than α, the null hypothesis is rejected. The data suggests that the model has utility for predicting discharge amount.

b $H_o: \beta_3 = 0$ $H_a: \beta_3 \neq 0$

$\alpha = 0.05$

The test statistic is: $t = \frac{b_3}{s_{b_3}}$ with df. = 6.

$$t = \frac{8.4}{199} = 0.04$$

P-value = 2(area under the 6 df t curve to the right of 0.04) $\approx 2(0.48) = 0.96$.

Since the P-value exceeds α, the null hypothesis is not rejected. The data suggests that the interaction term is not needed in the model, if the other two independent variables are in the model.

c   No. The model utility test is testing all variables simultaneously (that is, as a group). The t test is testing the contribution of an individual predictor when used in the presence of the remaining predictors. Results indicate that, given two out of the three predictors are included in the model, the third predictor may not be necessary.

14.47 The point prediction for mean phosphate adsorption when $x_1 = 160$ and $x_2 = 39$ is at the midpoint of the given interval. So the value of the point prediction is $(21.40 + 27.20)/2 = 24.3$. The t critical value for a 95% confidence interval is 2.23. The standard error for the point prediction is equal to $(27.20 - 21.40)/2(2.23) = 1.30$. The t critical value for a 99% confidence interval is 3.17. Therefore, the 99% confidence interval would be

$$24.3 \pm (3.17)(1.3) \Rightarrow 24.3 \pm 4.121 \Rightarrow (20.179, 28.421).$$

14.49 **a** $H_o: \beta_1 = \beta_2 = 0$

H_a: At least one of the two β_i's is not zero.

$\alpha = 0.05$

Test statistic: $F = \dfrac{SSRegr / k}{SSResid / [n - (k+1)]}$

$$F = \frac{237.52/2}{26.98/7} = 30.81$$

From Appendix Table VII, $0.001 >$ P-value.

Since the P-value is less than α, the null hypothesis is rejected. The data suggests that the fitted model is useful for predicting plant height.

b $\alpha = 0.05$. From the MINITAB output the t-ratio for b_1 is 6.57, and the t-ratio for b_2 is -7.69. The P-values for the testing $\beta_1 = 0$ and $\beta_2 = 0$ would be twice the area under the 7 df t curve to the right of 6.57 and 7.69, respectively. From Appendix Table IV, the
P-values are found to be practically zero. Both hypotheses would be rejected. The data suggests that both the linear and quadratic terms are important.

c   The point estimate of the mean y value when $x = 2$ is
$\hat{y} = 41.74 + 6.581(2) - 2.36(4) = 45.46$.

The 95% confidence interval is $45.46 \pm (2.37)(1.037) \Rightarrow 45.46 \pm 2.46$
$\Rightarrow (43.0, 47.92)$.

With 95% confidence, the mean height of wheat plants treated with
$x = 2$ ($10^2 = 100$ uM of Mn) is estimated to be between 43 and 47.92 cm.

d   The point estimate of the mean y value when $x = 1$ is
$\hat{y} = 41.74 + 6.58(1) - 2.36(1) = 45.96$.

The 90% confidence interval is $45.96 \pm (1.9)(1.031) \Rightarrow 45.96 \pm 1.96 \Rightarrow (44.0, 47.92)$.

With 90% confidence, the mean height of wheat plants treated with
$x = 1$ ($10 = 10$ uM of Mn) is estimated to be between 44 and 47.92 cm.

Section 14.4

14.51   One possible way would have been to start with the set of predictor variables consisting of all five variables, along with all quadratic terms, and all interaction terms. Then, use a selection procedure like backward elimination to arrive at the given estimated regression equation.

14.53   The model using the three variables x_3, x_9, x_{10} appears to be a good choice. It has an adjusted R^2 which is only slightly smaller than the largest adjusted R^2. This model is almost as good as the model with the largest adjusted R^2 but has two less predictors.

14.55   **a**   The model has 9 predictors.

H_o: $\beta_1 = \beta_2 = \beta_3 = \beta_4 = \beta_5 = \beta_6 = \beta_7 = \beta_8 = \beta_9 = 0$

H_a: at least one among β_1, β_2, β_3, β_4, β_5, β_6, β_7, β_8, β_9 is not zero

$\alpha = 0.05$ (for illustration).

$$F = \frac{R^2/k}{(1-R^2)/[n-(k+1)]}$$

n = 1856, $df_1 = k = 9$, $df_2 = n - (k + 1) = 1856 - 10 = 1846$.

$R^2 = = 0.3092$.

$$F = \frac{R^2/k}{(1-R^2)/[n-(k+1)]} = \frac{0.3092/9}{(1-0.3092)/1846} = 91.8$$

From Appendix Table VII, 0.001 > P-value.
Since the P-value is less than α, the null hypothesis is rejected. There does appear to be a useful linear relationship between ln(blood cadmium level) and at least one of the nine predictors.

b   If a backward elimination procedure was followed in the stepwise regression analysis, then the statements in the paper suggest that all variables except daily cigarette consumption and alcohol consumption were eliminated from the model. Of the two predictors left in the model, cigarette consumption would have a larger t-ratio than alcohol consumption.

There is an alternative procedure called the forward selection procedure which is available in most statistical software packages including MINITAB. According to this method one starts with a model having only the intercept term and enters one predictor at a time into the model. The predictor explaining most of the variance is entered first. The second predictor entered into the model is the one that explains most of the remaining variance, and so on. If the forward selection method was followed in the current problem then the statements in the paper would suggest that the variable to enter the model first is daily cigarette consumption and the next variable to enter the model is alcohol consumption. No further predictors were entered into the model.

14.57 **a**   Yes, they do show a similar pattern.

b   Standard error for the estimated coefficient of log of sales = (estimated coefficient)/t-ratio = 0.372/6.56 = 0.0567.

c   The predictor with the smallest (in magnitude) associated t-ratio is Return on Equity. Therefore it is the first candidate for elimination from the model. It has a t-ratio equal to 0.33 which is much less than $t_{out} = 2.0$. Therefore the predictor Return on Equity would be eliminated from the model if a backward elimination method is used with $t_{out} = 2.0$.

d   No. For the 1992 regression, the first candidate for elimination when using a backward elimination procedure is CEO Tenure since it has the smallest t-ratio (in magnitude).

e   We test H_0: Coefficient of Stock Ownership is equal to 0 versus H_a: Coefficient of Stock Ownership is less than 0. The t-ratio for this test is –0.58. Using Table IV from the appendix we find that the area to the left of the 153 d.f. t curve is approximately

0.3. So, the P-value for the test is approximately 0.3. Using MINITAB we find that the exact P-value is 0.2814.

14.59 From MINITAB, the best model with k variables and their summary statistics are

Number of Variables	Variables Included	R^2	Adjusted R^2	Cp
1	x_4	0.824	0.819	14.0
2	x_2, x_4	0.872	0.865	2.9
3	x_2, x_3, x_4	0.879	0.868	3.1
4	x_1, x_2, x_3, x_4	0.879	0.865	5.0

The best model, using the procedure of minimizing Cp, would use variables x_2, x_4. Hence, the set of predictor variables selected here is not the same as in problem **14.58**.

14.61 Using MINITAB, the best model with k variables has been found and summary statistics for each are given below.

k	Variables Included	R^2	Adjusted R^2	Cp
1	x_4	0.067	0.026	5.8
2	x_2, x_4	0.111	0.031	6.6
3	x_1, x_3, x_4	0.221	0.110	5.4
4	x_1, x_3, x_4, x_5	0.293	0.151	5.4
5	x_1, x_2, x_3, x_4, x_5	0.340	0.166	6.0

It appears that the model using x_1, x_3, x_4 is the best model, using the criterion of minimizing Cp.

Supplementary Exercises

14.63   **a** $H_o: \beta_1 = \beta_2 = \beta_3 = \ldots = \beta_{11} = 0$

H_a: at least one among β_i's is not zero

$\alpha = 0.01$

$$F = \frac{R^2/k}{(1-R^2)/[n-(k+1)]}$$

$n = 88, \ df_1 = k = 11, \ df_2 = n - (k+1) = 88 - 12 = 76$

$$F = \frac{R^2/k}{(1-R^2)/[n-(k+1)]} = \frac{.64/11}{(1-.64)/76} = \frac{.058182}{.004737} = 12.28$$

Appendix Table VII does not have entries for $df_1 = 11$, but using $df_1 = 10$ it can be determined that $0.001 > $ P-value.

14-18

Since the P-value is less than α, the null hypothesis is rejected. There does appear to be a useful linear relationship between y and at least one of the predictors.

b $\text{Adjusted R}^2 = 1 - \left[\dfrac{n-1}{n-(k+1)} \right] \dfrac{\text{SSResid}}{\text{SSTo}}$

To calculate adjusted R^2, we need the values for SSResid and SSTo. From the information given, we obtain:

$$s_e^2 = (5.57)^2 \Rightarrow 31.0249 = \frac{\text{SSResid}}{88-12} \Rightarrow \text{SSResid} = 76(31.0249) = 2357.8924$$

$$R^2 = .64 \Rightarrow .64 = 1 - \frac{\text{SSResid}}{\text{SSTo}} \Rightarrow .64 = 1 - \frac{2357.8924}{\text{SSTol}} \Rightarrow \frac{2357.8924}{\text{SSTo}} = .36$$

$$\Rightarrow SSTo = \frac{2357.8924}{.36} = 6549.7011.$$

So, $\text{Adjusted R}^2 = 1 - \left[\dfrac{n-1}{n-(k+1)} \right] \dfrac{\text{SSResid}}{\text{SSTo}} = 1 - \dfrac{87}{76} \left(\dfrac{2357.8924}{6549.7011} \right) = 1 - .4121 = .5879.$

c $\text{t-ratio} = \dfrac{b_1 - 0}{s_{b_1}} = 3.08 \Rightarrow s_{b_1} = \dfrac{b_1}{3.08} = \dfrac{.458}{3.08} = .1487$

$b_1 \pm (\text{t critical}) s_{b_1} \Rightarrow .458 \pm (2.00)(.1487) \Rightarrow .458 \pm .2974 \Rightarrow (.1606, .7554)$

From this interval, we estimate the value of β_1 to be between -0.1606 and 0.7554.

d   Many of the variables have t-ratios that are close to zero. The one with the smallest t in absolute value is x_9: certificated staff-pupil ratio. For this reason, I would eliminate x_9 first.

e H_o: $\beta_3 = \beta_4 = \beta_5 = \beta_6 = 0$

H_a: at least one among $\beta_3, \beta_4, \beta_5, \beta_6$ is non-zero.

None of the procedures presented in this chapter could be used. The two procedures presented tested "all variables as a group" or "a single variable's contribution".

14.65   **a**

Based on this scatterplot a quadratic model in x is suggested.

b $H_o: \beta_1 = \beta_2 = 0$

H_a: At least one of the two β_i's is not zero.

$\alpha = 0.05$

Test statistic: $F = \dfrac{SSRegr \,/\, k}{SSResid \,/\, [n-(k+1)]}$

$F = \dfrac{525.11/2}{61.77/5} = 21.25$

From Appendix Table VII, $0.01 > P\text{-value} > 0.001$.

Since the P-value is less than α, the null hypothesis is rejected. The data suggests that the quadratic model is useful for predicting glucose concentration.

c $H_o: \beta_2 = 0 \quad H_a: \beta_2 \neq 0$

$\alpha = 0.05$

The test statistic is: $t = \dfrac{b_2}{s_{b_2}}$ with df. = 5.

$t = \dfrac{1.7679}{.2712} = 6.52$

P-value = 2(area under the 5 df t curve to the right of 6.52) $\approx$ 2(0) = 0.

Since the P-value is less than α, the null hypothesis is rejected. The data suggests that the quadratic term cannot be eliminated from the model.

14.67 When n = 21 and k = 10, Adjusted $R^2 = 1 - 2(SSResid/SSTo)$.

Then Adjusted $R^2 < 0 \Rightarrow \frac{1}{2} < SSResid / SSTo = 1 - R^2 \Rightarrow 1/2 > R^2$.

Hence, when n = 21 and k = 10, Adjusted R^2 will be negative for values of R^2 less than 0.5.

14.69 First, the model using all four variables was fit. The variable age at loading (x_3) was deleted because it had the t-ratio closest to zero and it was between −2 and 2. Then, the model using the three variables x_1, x_2, and x_4 was fit. The variable time (x_4) was deleted because its t-ratio was closest to zero and was between −2 and 2. Finally, the model using the two variables x_1 and x_2 was fit. Neither of these variables could be eliminated since their t-ratios were greater than 2 in absolute magnitude. The final model then, includes slab thickness (x_1) and load (x_2). The predicted tensile strength for a slab that is 25 cm thick, 150 days old, and is subjected to a load of 200 kg for 50 days is $\hat{y} = 13 - 0.487(25) + 0.0116(200) = 3.145$.

14.71 **a**

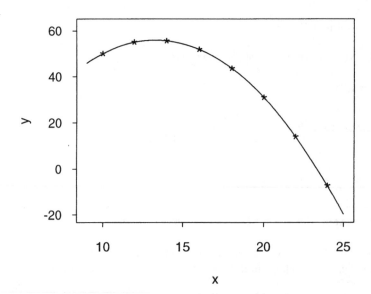

b The claim is very reasonable because 14 is close to where the smooth curve has its highest value.

14.73 **a** Output from MINITAB is given below.

```
The regression equation is:
Y = 1.56 + .0237 X1 - 0.000249 X2.

                   Coef          Stdev       t-ratio      p
Predictor
Constant         1.56450       0.07940       19.70      0.000
X1               0.23720       0.05556        4.27      0.000
X2              -0.00024908   0.00003205    -7.77      0.000

s = 0.05330         R-sq = 86.5%         R-sq(adj) = 85.3%

Analysis of Variance

SOURCE         DF     SS          MS         F        p
Regression      2    0.40151    0.20076    70.66    0.000
Error          22    0.06250    0.00284
Total          24    0.46402
```

b $H_o: \beta_1 = \beta_2 = 0$

H_a: At least one of the two β_i's is not zero.

$\alpha = 0.05$

Test statistic: $F = \dfrac{SSRegr / k}{SSResid / [n - (k+1)]}$

$F = \dfrac{.40151 / 2}{.0625 / 22} = 70.67$

From the Minitab output, the P-value associated with the F test is practically zero. Since the P-value is less than α, the null hypothesis is rejected.

c The value for R^2 is 0.865. This means that 86.5% of the total variation in the observed values for profit margin has been explained by the fitted regression equation. The value for s_e is 0.0533. This means that the typical deviation of an observed value from the predicted value is 0.0533, when predicting profit margin using this fitted regression equation.

d No. Both variables have associated t-ratios that exceed 2 in absolute magnitude. Hence, neither can be eliminated from the model.

e

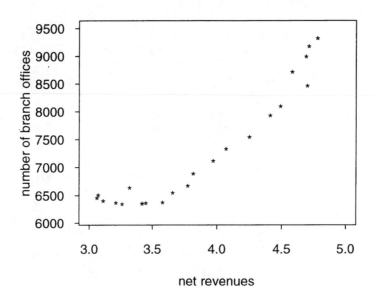

There do not appear to be any influential observations. However, there is substantial evidence of multicollinearity. The plot shows a pronounced linear relationship between x_1 and x_2. This is evidence of multicollinearity between x_1 and x_2.

Chapter 15
The Analysis of Variance

Section 15.1

15.1 **a** $0.01 > \text{P-value} > 0.001$

b $\text{P-value} > 0.10$

c $\text{P-value} = 0.01$

d $0.001 > \text{P-value}$

e $0.10 > \text{P-value} > 0.05$

f $0.05 > \text{P-value} > 0.01$ (Using $df_2 = 40$ and $df_2 = 60$ tables).

15.3 **a** Let μ_1, μ_2, μ_3 and μ_4 denote the true average length of stay in a hospital for health plans 1, 2, 3 and 4 respectively.

$H_o: \mu_1 = \mu_2 = \mu_3 = \mu_4$

$H_a:$ At least two of the four μ_i's are different.

b $df_1 = 4 - 1 = 3$ $df_2 = 32 - 4 = 28$ $\alpha = 0.01$

From Appendix Table VII, $0.05 > \text{P-value} > 0.01$, Since the P-value exceeds α, H_o is not rejected. Hence, it would be concluded that the average length of stay in the hospital is the same for the four health plans.

c $df_1 = 4 - 1 = 3$ $df_2 = 32 - 4 = 28$ $\alpha = 0.01$

From Appendix Table VII, $0.05 > \text{P-value} > 0.01$. Since the P-value exceeds α, H_o is not rejected. Therefore, the conclusion would be the same.

15.5   **a**   The required boxplot obtained using MINITAB is shown below. Price per acre values appear to be similar for 1996 and 1997 but 1998 values are higher. The mean price per acre values for each year are also plotted as a solid square within each box plot.

       b   Let μ_i denote the mean price per acre for vineyards in year i (i = 1, 2, 3).

$H_o: \mu_1 = \mu_2 = \mu_3$

H_a: At least two of the three μ_i's are different.

$\alpha = 0.01$

Test statistic: $F = \dfrac{MSTr}{MSE}$

$df_1 = k - 1 = 2 \quad df_2 = N - k = 15\text{-}3 = 12$.

$\bar{x}_1 = 35600, \bar{x}_2 = 36000, \bar{x}_3 = 43600$

$\bar{\bar{x}} = \dfrac{[5(35600) + 5(36000) + 5(43600)]}{(15)} = 38400$

$MSTr = [5(35600 - 38400)^2 + 5(36000 - 38400)^2 + 5(43600 - 38400)^2]/2$
$= 101600000$

$s_1 = 3847.077, s_2 = 3807.887, s_3 = 3911.521$
$MSE = [(5\text{-}1)(3847.077)^2 + (5\text{-}1)(3807.887)^2 + (5\text{-}1)(3911.521)^2]/12 = 14866667$

$F = \dfrac{MSTr}{MSE} = \dfrac{101600000}{14866667} = 6.83$

From Appendix Table VII, 0.05 > P-value > 0.01.

Since the P-value exceeds α, the null hypothesis is not rejected. At a significance level of $\alpha = 0.01$, the data does not support the claim that the true mean price per acre for the three years under consideration are different.

15.7   Let μ_i denote the mean level of chlorophyll concentration for plants in variety i (i = 1, 2, 3, 4).

$H_o: \mu_1 = \mu_2 = \mu_3 = \mu_4$

H_a: At least two of the four μ_i's are different.

$\alpha = 0.05$

15-2

Test statistic: $F = \dfrac{MSTr}{MSE}$

$df_1 = k - 1 = 3 \quad df_2 = N - k = 16.$

$\bar{\bar{x}} = \dfrac{[5(.3) + 5(.24) + 4(.41) + 6(.33)]}{(20)} = 0.316$

$MSTr = [5(0.3 - 0.316)^2 + 5(0.24 - 0.316)^2 + 4(0.41 - 0.316)^2 + 6(0.33 - 0.316)^2]/3$
$= 0.06668/3 = 0.022227$

$F = \dfrac{MSTr}{MSE} = \dfrac{.022227}{.013} = 1.71$

From Appendix Table VII, P-value > 0.10.

Since the P-value exceeds α, the null hypothesis is not rejected. The data does not suggest that true mean chlorophyll concentration differs for the four varieties.

15.9 μ_1 : true average importance rating of speed for owners of American cars.

μ_2 : true average importance rating of speed for owners of German cars.

μ_3 : true average importance rating of speed for owners of Japanese cars.

H_o: $\mu_1 = \mu_2 = \mu_3$

H_a: at least two among μ_1, μ_2, μ_3 are not equal.

$\alpha = 0.05$

$F = \dfrac{MSTr}{MSE}$

$df_1 = 2 \quad df_2 = (58 + 38 + 59) - 3 = 155 - 3 = 152$

$T = 58(3.05) + 38(2.87) + 59(2.67) = 443.49$

$\bar{\bar{x}} = \dfrac{T}{N} = \dfrac{443.49}{155} = 2.8612$

$SSTr = 58(3.05 - 2.8612)^2 + 38(2.87 - 2.8612)^2 + 59(2.67 - 2.8612)^2$
$= 2.066870 + 0.002925 + 2.157471 = 4.227267$

$MSTr = \dfrac{4.227267}{2} = 2.113634$

$$MSE = \frac{459.04}{152} = 3.02$$

$$F = \frac{MSTr}{MSE} = \frac{2.113634}{3.02} = .70$$

From Appendix Table VII, P-value > 0.10.

Since the P-value exceeds α, H_o is not rejected. It is plausible that the true average importance of speed rating is the same for the three groups.

15.11   Let μ_i denote the mean dry weight for concentration level i (i = 1, 2, ..., 10).

H_o: $\mu_1 = \mu_2 = ... = \mu_{10}$

H_a: At least two of the ten μ_i's are different.

$\alpha = 0.05$

Test statistic: $F = \dfrac{MSTr}{MSE}$

$df_1 = k - 1 = 9$ $df_2 = N - k = 30$.

$$F = \frac{MSTr}{MSE} = 1.895$$

From Appendix Table VII, 0.10 > P-value > 0.05.

Since the P-value exceeds α, the null hypothesis is not rejected. The data are consistent with the hypothesis that the true mean dry weight does not depend on the level of concentration.

15.13

Source of Variation	Degrees of Freedom	Sum of Squares	Mean Square	F
Treatments	3	75081.72	25027.24	1.70
Error	16	235419.04	14713.69	
Total	19	310500.76		

Let μ_i denote the mean number of miles to failure for brand i sparkplugs (i = 1, 2, 3, 4).

H_o: $\mu_1 = \mu_2 = \mu_3 = \mu_4$

H_a: At least two of the four μ_i's are different.

$\alpha = 0.05$

Test statistic: $F = \dfrac{MSTr}{MSE}$

$df_1 = k - 1 = 3$ $df_2 = N - k = 16$.   From the ANOVA table, $F = 1.70$.

From Appendix Table VII, P-value > 0.10.

Since the P-value exceeds α, the null hypothesis is not rejected. The data are consistent with the hypothesis that there is no difference between the mean number of miles to failure for the four brands of sparkplugs.

15.15   Computations: $\overline{\overline{x}} = [96(2.15) + 34(2.21) + 86(1.47) + 206(1.69)]/422$
$= 756.1/422 = 1.792$

$MSTr = [96(2.15 - 1.792)^2 + 34(2.21 - 1.792)^2 + 86(1.47 - 1.792)^2$
$+ 206(1.69 - 1.792)^2]/3 = 29.304/3 = 9.768$

$MSE = \dfrac{MSTr}{F} = \dfrac{9.768}{2.56} = 3.816$

Source of Variation	Degrees of Freedom	Sum of Squares	Mean Square	F
Treatments	3	29.304	9.768	2.56
Error	418	1595.088	3.816	
Total	421	1624.392		

Let μ_i denote the mean number of hours per month absent for employees of group i (i = 1, 2, 3, 4).

H_o: $\mu_1 = \mu_2 = \mu_3 = \mu_4$

H_a: At least two of the four μ_i's are different.

$\alpha = 0.01$

Test statistic: $F = \dfrac{MSTr}{MSE}$

$df_1 = k - 1 = 3$ $df_2 = N - k = 418$. From the ANOVA table, $F = 2.56$.

From Appendix Table VII, $0.10 > P\text{-value} > 0.05$.

Since the P-value exceeds α, the null hypothesis is not rejected. The data are consistent with the hypothesis that there is no difference between the mean number of hours per month absent for employees in the four groups.

15.17 Let μ_1, μ_2, and μ_3 denote the true mean fog indices for *Scientific America*, *Fortune*, and *New Yorker*, respectively.

$H_o: \mu_1 = \mu_2 = \mu_3$

H_a: At least two of the three μ_i's are different.

$\alpha = 0.01$

Test statistic: $F = \dfrac{MSTr}{MSE}$

$df_1 = k - 1 = 2$ $df_2 = N - k = 15$.

Computations: $\bar{\bar{x}} = 9.666$

Magazine	Mean	Standard Deviation
S.A.	10.968	2.647
F	10.68	1.202
N.Y.	7.35	1.412

$MSTr = [6(10.968 - 9.666)^2 + 6(10.68 - 9.666)^2 + 6(7.35 - 9.666)^2]/2 = 48.524/2 = 24.262$

$MSE = [(2.647)^2 + (1.202)^2 + (1.412)^2]/3 = 10.443/3 = 3.4812$

Source of Variation	Degrees of Freedom	Sum of Squares	Mean Square	F
Treatments	2	48.524	24.2620	6.97
Error	15	55.218	3.4812	
Total	17	100.742		

$F = \dfrac{MSTr}{MSE} = \dfrac{24.262}{3.4812} = 6.97$

From Appendix Table VII, $0.01 > \text{P-value} > 0.001$.

Since the P-value is less than α, the null hypothesis is rejected. The data suggests that there is a difference between at least two of the mean fog index levels for advertisements appearing in the three magazines.

Section 15.2

15.19 Since the intervals for $\mu_1 - \mu_2$ and $\mu_1 - \mu_3$ do not contain zero, μ_1 and μ_2 are judged to be different and μ_1 and μ_3 are judged to be different. Since the interval for $\mu_2 - \mu_3$ contains zero, μ_2 and μ_3 are judged not to be different. Hence, statement (iii) best describes the relationship between μ_1, μ_2, and μ_3.

15.21 μ_1 differs from μ_2; μ_1 differs from μ_3; μ_1 differs from μ_5; μ_2 differs from μ_4; μ_3 differs from μ_4; μ_4 differs from μ_5.

From the data given the following means were computed:

$$\bar{x}_1 = 16.35 \quad \bar{x}_2 = 11.63 \quad \bar{x}_3 = 10.5 \quad \bar{x}_4 = 14.96 \quad \bar{x}_5 = 12.3$$

Fabric	3	2	5	4	1
$\bar{x}$	10.5	11.63	12.3	14.96	16.35

15.23 $k = 4$ Error df $= (5 + 5 + 4 + 6) - 4 = 16$

From Appendix Table VIII, $q = 4.05$ for 95% confidence.

$$\mu_1 - \mu_2 : (.30 - .24) \pm 4.05 \sqrt{\frac{.013}{2}\left(\frac{1}{5} + \frac{1}{5}\right)} \Rightarrow .06 \pm .2065 \Rightarrow (-.1465, .2665)$$

$$\mu_1 - \mu_3 : (.30 - .41) \pm 4.05 \sqrt{\frac{.013}{2}\left(\frac{1}{5} + \frac{1}{4}\right)} \Rightarrow -.11 \pm .2190 \Rightarrow (-.3190, .1090)$$

$$\mu_1 - \mu_4 : (.30 - .33) \pm 4.05 \sqrt{\frac{.013}{2}\left(\frac{1}{5} + \frac{1}{6}\right)} \Rightarrow -.03 \pm .1977 \Rightarrow (-.2277, .1677)$$

$$\mu_2 - \mu_3 : (.24 - .41) \pm 4.05 \sqrt{\frac{.013}{2}\left(\frac{1}{5} + \frac{1}{4}\right)} \Rightarrow -.17 \pm .2190 \Rightarrow (-.3890, .0490)$$

$$\mu_2 - \mu_4 : (.24 - .33) \pm 4.05 \sqrt{\frac{.013}{2}\left(\frac{1}{5} + \frac{1}{6}\right)} \Rightarrow -.09 \pm .1977 \Rightarrow (-.2877, .1077)$$

$$\mu_3 - \mu_4 : (.41 - .33) \pm 4.05 \sqrt{\frac{.013}{2}\left(\frac{1}{4} + \frac{1}{6}\right)} \Rightarrow .08 \pm .2108 \Rightarrow (-.1308, .2908)$$

There appear to be no pairwise differences.

Variety	2. RO	1. BI	4. TO	3. WA
Mean	0.24	0.30	0.33	0.41

No differences are detected. This conclusion is in agreement with the results of the F test done in problem 15.7.

15.25

Group	Simultaneous	Sequential	Control
Mean	$\bar{x}_3$	$\bar{x}_2$	$\bar{x}_1$

15.27 The mean water loss when exposed to 4 hours fumigation is different from all other means. The mean water loss when exposed to 2 hours fumigation is different from that for levels 16 and 0, but not 8. The mean water losses for duration 16, 0, and 8 hours are not different from one another. No other differences are significant.

15.29 **a** H_o: $\mu_1 = \mu_2 = \mu_3 = \mu_4 = \mu_5$

H_a: At least two of the five μ_i's are different.

$\alpha = 0.05$

Test statistic: $F = \dfrac{MSTr}{MSE}$

$df_1 = k - 1 = 4$ $df_2 = N - k = 15.$

From the data, the following statistics were calculated.

Hormone	1	2	3	4	5
n	4	4	4	4	4
mean	12.75	17.75	17.50	11.50	10.00
Total	51	71	70	46	40
variance	17.583	12.917	8.333	21.667	15.333

$$\bar{\bar{x}} = \frac{T}{N} = \frac{278}{20} = 13.9$$

$$\begin{aligned} SSTr &= 4(12.75 - 13.9)^2 + 4(17.75 - 13.9)^2 + 4(17.50 - 13.9)^2 + 4(11.5 - 13.9)^2 \\ &\quad + 4(10 - 13.9)^2 \\ &= 5.29 + 59.29 + 51.84 + 23.04 + 60.840 = 200.3 \end{aligned}$$

$$SSE = 3(17.583) + 3(12.917) + 3(8.333) + 3(21.667) + 3(15.333) = 227.499$$

$$df_1 = k - 1 = 5 - 1 = 4 \quad df_2 = (4 + 4 + 4 + 4 + 4) - 5 = 15$$

$$MSTr = \frac{SSTr}{k-1} = \frac{200.3}{4} = 50.075$$

$$MSE = \frac{SSE}{N-k} = \frac{227.499}{15} = 15.1666$$

$$F = \frac{MSTr}{MSE} = \frac{50.075}{15.1666} = 3.30$$

From Appendix Table VII, $0.05 > P\text{-value} > 0.01$.

Since the P-value is less than α, H_o is rejected. The data supports the conclusion that the mean plant growth is not the same for all five growth hormones.

b $k = 5$ Error df $= 15$

From Appendix Table VIII, $q = 4.37$.

Since the sample sizes are the same, the $\pm$ factor is the same for each comparison.

$$4.37\sqrt{\frac{15.1666}{2}\left(\frac{1}{4} + \frac{1}{4}\right)} = 8.51$$

$\mu_1 - \mu_2$: $(12.75 - 17.75) \pm 8.51 \Rightarrow -5 \pm 8.51 \Rightarrow (-13.51, 3.51)$

$\mu_1 - \mu_3$: $(12.75 - 17.50) \pm 8.51 \Rightarrow -4.75 \pm 8.51 \Rightarrow (-13.26, 3.76)$

$\mu_1 - \mu_4$: $(12.75 - 11.5) \pm 8.51 \Rightarrow 1.25 \pm 8.51 \Rightarrow (-7.26, 9.76)$

$\mu_1 - \mu_5$: $(12.75 - 10) \pm 8.51 \Rightarrow 2.75 \pm 8.51 \Rightarrow (-5.76, 11.26)$

$\mu_2 - \mu_3$: $(17.75 - 17.50) \pm 8.51 \Rightarrow 0.25 \pm 8.51 \Rightarrow (-8.26, 8.76)$

$\mu_2 - \mu_4$: $(17.75 - 11.5) \pm 8.51 \Rightarrow 6.25 \pm 8.51 \Rightarrow (-2.26, 14.76)$

$\mu_2 - \mu_5$: $(17.75 - 10) \pm 8.51 \Rightarrow 7.75 \pm 8.51 \Rightarrow (-0.76, 16.26)$

$\mu_3 - \mu_4$: $(17.5 - 11.5) \pm 8.51 \Rightarrow 6.0 \pm 8.51 \Rightarrow (-2.51, 14.51)$

$\mu_3 - \mu_5$: $(17.5 - 10) \pm 8.51 \Rightarrow 7.5 \pm 8.51 \Rightarrow (-1.01, 16.01)$

$\mu_4 - \mu_5$: $(11.5 - 10) \pm 8.51 \Rightarrow 1.5 \pm 8.51 \Rightarrow (-7.01, 10.01)$

No significant differences are determined using the T-K method.

Section 15.3

15.31 a

Source of Variation	Degrees of Freedom	Sum of Squares	Mean Square	F
Treatments	2	11.7	5.85	0.37
Blocks	4	113.5	28.375	
Error	8	125.6	15.7	
Total	14	250.8		

b

H_o: The mean appraised value does not depend on which assessor is doing the appraisal.

H_a: The mean appraised value does depend on which assessor is doing the appraisal.

$\alpha = 0.05$

Test statistic: $F = \dfrac{MSTr}{MSE}$

$df_1 = k - 1 = 2$ $df_2 = (k-1)(l-1) = 8$. From the ANOVA table, $F = 0.37$.

From Appendix Table VII, P-value > 0.10.

Since the P-value exceeds α, the null hypothesis is not rejected. The mean appraised value does not seem to depend on which assessor is doing the appraisal.

15.33

Source of Variation	Degrees of Freedom	Sum of Squares	Mean Square	F
Treatments	2	3.97	1.985	79.4
Blocks	7	0.2503	0.0358	
Error	14	0.3497	0.025	
Total	23	4.57		

H_o: The mean energy use does not depend on the type of oven.

H_a: The mean energy use does depend on the type of oven.

$\alpha = 0.01$

Test statistic: $F = \dfrac{MSTr}{MSE}$

$df_1 = k - 1 = 2$ and $df_2 = (k - 1)(l - 1) = 14$. From the ANOVA table, $F = 79.4$.

From Appendix Table VII, $0.001 > $ P-value.

Since the P-value is less than α, the null hypothesis is rejected. The data suggests quite strongly that the mean energy use depends on the type of oven used.

15.35 **a** Other environmental factors (amount of rainfall, number of days of cloudy weather, average daily temperature, etc.) vary from year to year. Using a randomized complete block helps control for variation in these other factors.

b $SSTr = 3[(138.33 - 149)^2 + (152.33 - 149)^2 + (156.33 - 149)^2] = 536.00$

$SSBl = 3[(173 - 149)^2 + (123.67 - 149)^2 + (150.33 - 149)^2] = 3658.13$

Source of Variation	Degrees of Freedom	Sum of Squares	Mean Square	F
Treatments	2	536.00	268.00	14.91
Blocks	2	3658.13	1829.07	
Error	4	71.87	17.97	
Total	8	4266.00		

H_o: The mean height does not depend on the rate of application of effluent.

H_a: The mean height does depend on the rate of application of effluent.

$\alpha = 0.05$

Test statistic: $F = \dfrac{MSTr}{MSE}$

$df_1 = k - 1 = 2$ and $df_2 = (k - 1)(l - 1) = 4$. From the ANOVA table, $F = 14.91$.

From Appendix Table VII, $0.05 > $ P-value > 0.01.

Since the P-value is less than α, the null hypothesis is rejected. The data suggests that the mean height of cotton plants does depend on the rate of application of effluent.

c   Since the sample sizes are equal, the ± factor is the same for each comparison.

$k = 3$ Error df $= 4$ $q = 5.04$ (This value came from a more extensive table of critical values for the Studentized range distribution than the text provides.)

The ± factor is: $5.04\sqrt{\dfrac{17.97}{3}} = 12.33$

Application Rate	1	2	3
Mean	138.33	152.33	156.33

The mean height under an application rate of 350 differs from those using application rates of 440 and 515.

The mean height using application rates of 440 and 515 do not differ.

15.37 H_o: The mean height does not depend on the seed source.

H_a: The mean height does depend on the seed source.

$\alpha = 0.05$

Test statistic: $F = \dfrac{MSTr}{MSE}$

Source of Variation	Degrees of Freedom	Sum of Squares	Mean Square	F
Treatments	4	4.543	1.136	0.868
Blocks	3	7.862	2.621	
Error	12	15.701	1.308	
Total	19	28.106		

$df_1 = k - 1 = 4$, $df_2 = (k - 1)(l - 1) = 12$, and $F = 0.868$.

From Appendix Table VII, P-value > 0.10.
Since the P-value exceeds α, the null hypothesis is not rejected. The data are consistent with the hypothesis that mean height does not depend on the seed source.

Section 15.4

15.39 **a**

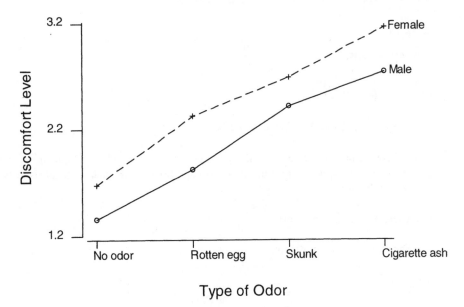

b   The graphs for males and females are very nearly parallel. There does not appear to be an interaction between gender and type of odor.

15.41 **a**   The plot does suggest an interaction between peer group and self-esteem. The change in average response, when changing from low to high peer group, is not the same for the low self-esteem group and the high self-esteem group. This is indicated by the non-parallel lines.

b   The change in the average response is greater for the low self-esteem group than it is for the high self-esteem group, when changing from low to high peer group interaction. Therefore, the authors are correct in their statement.

15.43   Analysis of Variance:

Source of Variation	Degrees of Freedom	Sum of Squares	Mean Square	F
Size (A)	2	0.088	0.044	4.00
Species (B)	1	0.048	0.048	4.363
Size by Species	2	0.048	0.024	2.18
Error	12	0.132	0.011	
Total	17	0.316		

H_o: There is no interaction between Size (A) and Species (B).

H_a: There is interaction between Size and Species.

$\alpha = 0.01$

The test statistic is: $F_{AB} = \dfrac{MSAB}{MSE}$.

$df_1 = 2$, $df_2 = 12$, and $F_{AB} = 2.18$

From Appendix Table VII, P-value > 0.10.

Since the P-value exceeds α, the null hypothesis is not rejected. The data are consistent with the hypothesis of no interaction between Size and Species. Hence, hypothesis tests on main effects will be done.

H_o: There are no size main effects.

H_a: There are size main effects.

$\alpha = 0.01$

The test statistic is: $F_A = \dfrac{MSA}{MSE}$.

$df_1 = 2$, $df_2 = 12$, and $F_A = 4.00$.

From Appendix Table VII, $0.05 > $ P-value $ > 0.01$.

Since the P-value exceeds α, the null hypothesis of no size main effects is not rejected. When using $\alpha=0.01$, the data support the conclusion that there are no differences between the mean preference indices for the three sizes of bass.

H_o: There are no species main effects.

H_a: There are species main effects.

$\alpha = 0.01$

The test statistic is: $F_B = \dfrac{MSB}{MSE}$.

$df_1 = 1$, $df_2 = 12.$, and $F_B = 4.363$

From Appendix Table VII, P-value > 0.05.

Since the P-value exceeds α, the null hypothesis of no species main effects is not rejected. At a significance level of $\alpha = 0.01$, the data are consistent with the hypothesis that there are no differences between the mean preference indices for the three species of bass.

15.45 **a** H_0: There is no interaction between Greek status (A) and Year (B).

H_a: There is interaction between Greek status and year in college.

$\alpha = 0.05$ (for illustration)

The test statistic is: $F_{AB} = \dfrac{MSAB}{MSE}$.

Factor A (Greek status) has 2 levels, so $df_A = 1$.

Factor B (Year in college) has 4 levels, so $df_B = 3$.

Hence $df_{AB} = (1)(3) = 3$.

Since the total number of observations is $75 + 57 = 132$, the d.f. for Total is 131 and the d.f. for Error is $131-(1+3+3) = 124$.

Hence, for the test of interaction we have

$df_1 = 3$, $df_2 = 124$, $F_{AB} = 0.70$.

From Appendix Table VII, P-value > 0.10.

Since the P-value is greater than α, the null hypothesis is not rejected. The data do not suggest existence of an interaction between Greek status and year in college. (Hence tests about main effects may be performed.)

b H_0: There is no Greek status main effect.

H_a: There is a Greek status main effect.

$\alpha = 0.05$ (for illustration)

The test statistic is: $F_A = \dfrac{MSA}{MSE}$.

$df_1 = 1$, $df_2 = 124$, and $F_A = 20.53$

From Appendix Table VII, $0.001 > $ P-value.

Since the P-value is smaller than α, the null hypothesis of no Greek status main effect is rejected. The data provide evidence to conclude that there is a difference between the mean "Self-esteem" scores of students belonging to a fraternity and those who do not belong to a fraternity.

c H_0: There are no year main effects.

H_a: There are year main effects.

$\alpha = 0.05$ (for illustration)

The test statistic is: $F_B = \dfrac{MSB}{MSE}$.

$df_1 = 3$, $df_2 = 124$, and $F_B = 2.59$

From Appendix Table VII, $0.10 > $ P-value > 0.05.

Since the P-value is greater than α, the null hypothesis of no year main effects is rejected. The data do not suggest the existence of differences among the mean "Self-esteem" scores of students belonging to different years in college.

15.47 The test for no interaction would have $df_1 = 2$, $df_2 = 120$, and $F_{AB} < 1$. From Appendix Table VII, P-value > 0.10. Since the P-value exceeds the α of 0.05, the null hypothesis of no interaction is not rejected. Since there appears to be no interaction, hypothesis tests on main effects are appropriate.

The test for no A main effects would have $df_1 = 2$, $df_2 = 120$, and $F_A = 4.99$. From Appendix Table VII, $0.05 > $ P-value > 0.01. Since the P-value is less than α, the null hypothesis of no A main effects is rejected. The data suggests that the expectation of opportunity to cheat affects the mean test score.

The test for no B main effects would have $df_1 = 1$, $df_2 = 120$, and $F_B = 4.81$. From Appendix Table VII, $0.05 > $ P-value > 0.01. Since the P-value is less than α, the null hypothesis of no B main effects is rejected. The data suggests that perceived payoff affects the mean test score.

15.49 **a**

Source of Variation	Degrees of Freedom	Sum of Squares	Mean Square	F
Race	1	857	857	5.57
Sex	1	291	291	1.89
Race by Sex	1	32	32	0.21
Error	36	5541	153.92	
Total	39	6721		

H_0: There is no interaction between race and sex.

H_a: There is interaction between race and sex.

$\alpha = 0.01$

15-16

Test statistic: $F_{AB} = \dfrac{MSAB}{MSE}$

$df_1 = 1$, $df_2 = 36$, and $F_{AB} = 0.21$. From Appendix Table VII, P-value > 0.10.

Since the P-value exceeds α, the null hypothesis of no interaction between race and sex is not rejected. Thus, hypothesis tests for main effects are appropriate.

b H_o: There are no race main effects.

H_a: There are race main effects.

$\alpha = 0.01$

Test statistic: $F_A = \dfrac{MSA}{MSE}$

$df_1 = 1$, $df_2 = 36$, and $F_A = 5.57$. From Appendix Table VII, $0.05 > $ P-value > 0.01

Since the P-value exceeds α, the null hypothesis of no race main effects is not rejected. The data are consistent with the hypothesis that the true average lengths of sacra do not differ for the two races.

c H_o: There are no sex main effects.

H_a: There are sex main effects.
$\alpha = 0.01$

Test statistic: $F_B = \dfrac{MSB}{MSE}$

$df_1 = 1$, $df_2 = 36$, and $F_B = 1.89$. From Appendix Table VII, P-value > 0.10.

Since the P-value exceeds α, the null hypothesis of no sex main effects is not rejected. The data are consistent with the hypothesis that the true average lengths of sacra do not differ for males and females.

15.51   The following Anova table was obtained using the statistical software package Minitab.

Source of Variation	Degrees of Freedom	Sum of Squares	Mean Square	F
Rate	2	469.70	234.85	76.14
Soil Type	2	333.94	166.97	54.14
Error	4	12.34	3.08	
Total	8	815.98		

For each test, $df_1 = 2$, $df_2 = 4$. From Appendix Table VII, the P-value for the rate test is less than 0.001 and the P-value for the soil type is between 0.01 and 0.001. Since the P-values are less than $\alpha = 0.01$, both null hypotheses are rejected using a 0.01 level of significance. It appears that the total phosphorus uptake depends upon application rate as well as soil types.

$k = 3$, Error $df = 4$, $q = 5.04$ (This value came from a more extensive table of critical values for the Studentized range distribution than the text provides.)

The $\pm$ factor $= 5.04\sqrt{\dfrac{3.08}{3}} = 5.107$.

Soil Type	Ramiha	Konini	Wainui
Mean	10.217	20.957	24.557

The mean effect of soil types Komini and Wainui on total phosphorus uptake is the same, but differs for soil type Ramiha.

Supplementary Exercises

15.53

Source of Variation	Degrees of Freedom	Sum of Squares	Mean Square	F
Treatments	11	592.66	53.878	4.13
Error	175	2282.98	13.046	
Total	186	2875.64		

From Appendix Table VII, $0.001 > $ P-value. At $\alpha = 0.001$, H_o is rejected. There is strong sample evidence to support the conclusion that the average audience age is not the same for magazines in which advertisements appeared.

15.55 $H_o: \mu_1 = \mu_2 = \mu_3 = \mu_4$

H_a: at least two among μ_1, μ_2, μ_3, μ_4 are not equal.

$\alpha = 0.05$

$F = \dfrac{MSTr}{MSE}$

$N = 52$, $df_1 = 3$ $df_2 = 48$

$SSTr = SSTo - SSE = 682.10 - 506.19 = 175.91$

$$\text{MSTr} = \frac{175.91}{3} = 58.637$$

$$\text{MSE} = \frac{506.19}{48} = 10.546$$

$$F = \frac{\text{MSTr}}{\text{MSE}} = \frac{58.637}{10.546} = 5.56$$

From Appendix Table VII, $0.01 > \text{P-value} > 0.001$.

Since the P-value is less than α, the null hypothesis is rejected. The sample evidence supports the conclusion that the mean grievance rate is not the same for the four groups.

Since the sample sizes are equal, the $\pm$ factor is the same for each comparison.

$k = 4$, Error df $= 48$. From Appendix Table VIII, $q \approx 3.79$.

$$\pm \text{factor} = 3.79\sqrt{\frac{10.546}{13}} = 3.414$$

Group	Apathetic	Conservative	Erratic	Strategic
Mean	2.96	4.91	5.05	8.74

The mean grievance rate for the strategic group is larger than the mean grievance rate for the other three groups. No other significant differences appear to be present.

15.57   **a** H_o: $\mu_1 = \mu_2 = \mu_3 = \mu_4$

H_a: at least two among μ_1, μ_2, μ_3, μ_4 are not equal.

$\alpha = 0.05$

$$F = \frac{\text{MSTr}}{\text{MSE}}$$

From the data the following summary statistics were computed:

$n_1 = 6$, $\bar{x}_1 = 4.923$, $s_1^2 = .000107$ $n_2 = 6$, $\bar{x}_2 = 4.923$, $s_2^2 = .000067$

$n_3 = 6$, $\bar{x}_3 = 4.917$, $s_3^2 = .000147$ $n_4 = 6$, $\bar{x}_4 = 4.902$, $s_4^2 = .000057$

$$T = 6(4.923) + 6(4.923) + 6(4.917) + 6(4.902) = 117.99$$

$$\bar{\bar{x}} = \frac{117.99}{24} = 4.916$$

$$SSTr = 6(4.923 - 4.916)^2 + 6(4.923 - 4.916)^2 + 6(4.917 - 4.916)^2$$
$$+ 6(4.902 - 4.916)^2$$
$$= 0.000294 + 0.000294 + 0.000006 + 0.001176 = 0.001770$$

$$SSE = 5(0.000107) + 5(0.000067) + 5(0.000147) + 5(0.000057) = 0.001890$$

$$MSTr = \frac{SSTr}{k-1} = \frac{.001770}{3} = .00059$$

$$MSE = \frac{SSE}{N-k} = \frac{.001890}{24-4} = .0000945$$

$$F = \frac{MSTr}{MSE} = \frac{.00059}{.0000945} = 6.24$$

$df_1 = 3$, $df_2 = 20$. From Appendix Table VII, $0.01 > $ P-value > 0.001.

Since the P-value is less than α, the null hypothesis is rejected. The sample data supports the conclusion that there are differences in the true average iron content for the four storage periods.

b $k = 4$, Error df = 20. From Appendix Table VIII, $q = 3.96$. Since the sample sizes are equal, the $\pm$ factor is the same for each comparison.

$$\pm \text{ factor} = 3.96\sqrt{\frac{.0000945}{6}} = .0157$$

Storage Period	4	3	2	1
Mean	4.902	4.917	4.923	4.923

The mean for storage period 4 differs from the means for storage periods 2 and 1. No other significant differences are present.

15.59   The following ANOVA table was produced by using the computer packaged MINITAB. Because MINITAB does not compute F ratios, the user must do so from the mean squares which are given. The F ratios are found by the user dividing the appropriate mean squares.

SOURCE	DF	SS	MS	F
Oxygen	3	0.1125	0.0375	2.76
Sugar	1	0.1806	0.1806	13.28
Interaction	3	0.0181	0.0060	0.44
Error	8	0.1087	0.0136	
Total	15	0.4200		

Test for interaction: $\alpha = 0.05$

$df_1 = 3$, $df_2 = 8$. The F ratio to test for interaction is $F_{AB} = 0.44$. From Appendix Table VII, P-value > 0.10. Since this P-value exceeds α, the null hypothesis of no interaction is not rejected. Thus, it is appropriate to test for main effects.

Test for oxygen main effects: $\alpha = 0.05$

$df_1 = 3$ and $df_2 = 8$. The F ratio to test for oxygen main effects is $F_A = 2.76$. From Appendix Table VII, P-value > 0.10. Since the P-value exceeds α, the null hypothesis of no oxygen main effects is not rejected. The data are consistent with the hypothesis that the true average ethanol level does not depend on which oxygen concentration is used.

Test for sugar main effects: $\alpha = 0.05$

$df_1 = 1$ and $df_2 = 8$. The F ratio to test for sugar main effects is $F_B = 13.28$. From Appendix Table VII, $0.01 >$ P-value > 0.001. Since the P-value is less than α, the null hypothesis of no sugar main effects is rejected. The data suggests that the true average ethanol level does differ for the two types of sugar.

15.61

Source of Variation	Degrees of Freedom	Sum of Squares	Mean Square	F
A main effects	1	322.667	322.667	980.75
B main effects	3	35.623	11.874	36.09
Interaction	3	8.557	2.852	8.67
Error	16	5.266	0.329	
Total	23	372.113		

Test for interaction: $\alpha = 0.05$

$df_1 = 3$ and $df_2 = 16$, and from the Anova table $F_{AB} = 8.67$. From Appendix Table VII, $0.01 >$ P-value > 0.001. Since the P-value is less than α, the null hypothesis of no interaction is rejected. The data suggests that there is interaction between mortar type and submersion period. Thus, no tests for main effects will be performed.

15.63   Let μ_1, μ_2, and μ_3 denote the mean lifetime for brands 1, 2 and 3 respectively.

H_o: $\mu_1 = \mu_2 = \mu_3$

H_a: At least two of the three μ_i's are different.

$\alpha = 0.05$

Test statistic: $F = \dfrac{MSTr}{MSE}$

Computations: $\Sigma x^2 = 45171$, $\bar{\bar{x}} = 46.1429$, $\bar{x}_1 = 44.571$, $\bar{x}_2 = 45.857$, $\bar{x}_3 = 48$

$SSTo = 45{,}171 - 21(46.1429)^2 = 45171 - 44712.43 = 458.57$

$SSTr = 7(44.571)^2 + 7(45.857)^2 + 7(48)^2 - 44712.43 = 44754.43 - 44712.43 = 42$

$SSE = SSTo - SSTr = 458.57 - 42 = 416.57$

Source of Variation	Degrees of Freedom	Sum of Squares	Mean Square	F
Treatments	2	42	21	0.907
Error	18	416.57	23.14	
Total	20	458.57		

$df_1 = 2$ and $df_2 = 18$. The F ratio is 0.907. From Appendix Table VII, P-value > 0.10. Since the P-value exceeds α, H_o is not rejected at level of significance 0.05. The data are consistent with the hypothesis that there are no differences in true mean lifetimes of the three brands of batteries.

15.65 The transformed data is:

						mean
Brand 1	3.162	3.742	2.236	3.464	2.828	3.0864
Brand 2	4.123	3.742	2.828	3.000	3.464	3.4314
Brand 3	3.606	4.243	3.873	4.243	3.162	3.8254
Brand 4	3.742	4.690	3.464	4.000	4.123	4.0038

$SSTo = 264.001 - 20(3.58675)^2 = 264.001 - 257.2955 = 6.7055$

$SSTr = [5(3.0864)^2 + 5(3.4314)^2 + 5(3.8254)^2 + 5(4.0038)^2] - 257.2955$
$\quad = 259.8409 - 257.2955 = 2.5454$

$SSE = 6.7055 - 2.5454 = 4.1601$

Let μ_1, μ_2, μ_3, μ_4 denote the mean of the square root of the number of flaws for brand 1, 2, 3 and 4 of tape, respectively.

$H_o: \mu_1 = \mu_2 = \mu_3 = \mu_4$

H_a: At least two of the four μ_i's are different.

$\alpha = 0.01$

$$F = \frac{MSTr}{MSE}$$

$df_1 = 3$ and $df_2 = 16$.

$$F = \frac{2.5454 / 3}{4.1601 / 16} = 3.26$$

From Appendix Table VII, P-value > 0.01.

Since the P-value exceeds α, H_0 is not rejected. The data are consistent with the hypothesis that there are no differences in true mean square root of the number of flaws for the four brands of tape.